To Provide *and* Maintain a Navy

WHY NAVAL PRIMACY IS AMERICA'S FIRST, BEST STRATEGY

JERRY HENDRIX

FOCSLE

Focsle LLP
Annapolis, MD 21401

Hendrix, Henry J.
To Provide and Maintain a Navy: Why Naval Primacy is America's First, Best Strategy

ISBN 978-0-9600391-9-7

Printed in the United States of America

Dedication

For my daughters, Amanda and Michaela,
"Dad" is the best title I ever had.

"To Provide and Maintain a Navy"
Why Naval Primacy Is America's First, Best Strategy

Jerry Hendrix

Contents

Foreword

by John Lehman,
Former Secretary of the Navy

As Dr. (and retired Navy captain) Jerry Hendrix lays out so well in this timely and incisive analysis, history has certainly moved on from our late twentieth-century victories in the Cold War and Operation Desert Storm. New threats to our country, its prosperity, and way of life have clearly emerged now from what he aptly terms the "second interwar period." We are now squarely in the midst of a new great power competition — indeed, a "Second Cold War" — with China and Russia, while also confronting other hostile actors in Iran and North Korea, and Middle Eastern terrorists as well. Meanwhile, as those threats have been growing, we and our allies have cut back far too drastically on our own defense capabilities, especially at sea. Force levels have been slashed, budgets cut, exercises trimmed or cancelled, and training and maintenance pared back. As a result, deterrence of our adversaries, reassurance of our allies and partners, and sustainment and protection of the successful global political, economic, and military system instituted after World War II have all deteriorated.

It does not have to be that way. We in the Navy faced a parallel situation forty years ago, but by mobilizing the financial, technological, conceptual, and military resources of our country and its allies, we developed a coherent and realistic strategy, achievable and balanced force goals, and the hard-headed, competitive procurement measures that were necessary to push back successfully against the Soviets at sea. In doing so, our morale rose to winning heights.

The Chinese and Russians of 2020 are not, of course, clones of the Soviets of 1981. But the broad outline of their expanding global threat is similar, and similar as well to the 1940 Axis agreement between Nazi Germany, Italy, and Japan. Iran has now joined the China-Russia anti-US axis and now regularly participates in joint military and naval training exercises with them. Threats grow before our eyes.

While funding still lags, America's material and intellectual resources today remain strong — indeed preeminent — but they do need to be focused, mobilized, directed, and deployed correctly. First, we need a *strategy* — clear and unequivocal. In keeping with the nature of the US naval service, it needs to be aggressive, offensive, fast-reacting, lethal, sustainable, global, forward, joint when it matters, and allied. We developed such a strategy in the

1980s — the Forward Maritime Strategy — that described what we were about in peacetime and in the face of crises, and what we planned to carry out should deterrence fail and warfighting at sea and from the sea commence. The strategy we need now (as then) must be informed by the very best intelligence — including views from outside the Intelligence Community — as to what our adversaries have in the way of capabilities, as well as goals and intentions. The strategy must not be a replay of the Cold War. It must be directed at the extreme vulnerability of the economies and commerce of the Axis, and not pursue the strategy that they expect from us and are building their forces to defeat. The strategy needs to be specific enough so that operational commanders and resource sponsors alike are able to figure out what to keep and use, what to discard, and what to build. We know how to do this. We've done it before. We can do it again, and then broadcast it far and wide — with appropriate security caveats — so that it too becomes the dominant tool of deterrence, reassurance, global economic system maintenance, and warfighting.

The strategy should provide broad guidance on what it seeks to accomplish — to deter hostilities in the first place, but if that fails, to protect US and allied forces, sink enemy combat and merchant fleets, selectively attack targets ashore using both kinetic and non-kinetic means, blockade and mine their harbors and seaports, and interdict and shut down their commerce from air, submarine, surface ships, and some of the 25,000 islands in the Pacific that cannot be pretargeted by the Axis. These islands will soon be unpredictably accessible by the brilliant new configuration and capabilities of the US Marines integrated in the modernizing US Navy and its global allies. The strategy should make clear that cover and deception will be key to its success (i.e., if they don't know where we are, they can't hit us). The strategy should be executable at longer ranges than the maritime strategy of the 1980s was, given the development and deployment over the past few decades of modern Russian and Chinese long-range missiles and aircraft, in part in reaction to the US Maritime Strategy of the '80s.

The strategy should be robust, flexible, and prepared for unexpected breakthroughs and setbacks, unanticipated threats or domestic consideration, neutrals who join us, allies who don't, and so on.

So, strategy first. Then a realistic, sufficient, and achievable *force goal*, each element of which is derived from the strategy. For us it was 600 ships, including 15 carrier battle groups, 100 attack submarines, 100 frigates, and more, capable of carrying out all the Navy's anticipated warfare areas identified in the strategy: strike, amphibious, anti-air, anti-submarine, electronic warfare, blockade, and the like. The 600-ship goal was the result of a decade of discussions, analyses, exercises, and war games. Lots of alternatives were examined — we got a lot of "help" on that from outside the Navy — but we thought our numbers were solid enough to enable us to back them consistently, year after year, giving operational commanders, program managers, and contractors sufficient stability to plan, build, and equip the force rapidly and efficiently. Tweaking the force goal year after year while chasing every new shiny object brought forth by everyone from warfighters to futurists to scientists to politicians would have been ruinous.

As Dr. Hendrix points out, carriers will remain essential in the US fleet, but they won't necessarily be the super-carriers of the maritime strategy and the 600-ship Navy. Navy decision-makers should examine, for example, the virtues for the twenty-first century of Midway-class-sized diesel/gas turbine carriers, with adequate protection against hypersonic and ballistic missile hits, as well as a new-technology, non-nuclear submarine — lethal and affordable in large numbers. And given the longer ranges needed to execute the strategy, the force goal should give priority to new long-range carrier aircraft with new long-range missiles, just as it had in the 1970s and '80s. The Navy will also need Dr. Hendrix's recommended 456 ship fleet of frigates, corvettes, and smaller amphibious ships, to patrol and police the world oceans but easily integrate into fleet warfighting formations should the need arise.

Next, the program to carry out the strategy needs to be *affordable*. Whatever the force goal required, it will cost a lot of money. The Navy must demonstrate to the American people and the Congress that they are getting the most bang for their bucks. The way to do that is clear: we did it in the 1980s. Central to the effort must be revitalizing cost-cutting competition, and ending the practice of awarding sole-source contracts to monopolies, who tend to be entrenched opponents of innovation and cost-control.

Ship designs should be such that they can be competed, year after year, between at least two yards, not awarded to only one entitled and stodgy builder. Service lives should be extended — for example, for the most modernized latest Los Angeles–class attack submarines. Constant design changes and gold-plating need to be rigorously opposed, with hard-nosed procedures in place to do so automatically and routinely.

While demonstrating its affordability bona fides, the Navy — and proponents of increased US naval power in politics, academia, analysis, and industry — should argue for a larger slice of the defense budget pie, doubtless necessarily at the expense of US ground forces and the so-called "fourth estate." The strategy will demand it and the appropriate force goal will require it, even with strict affordability measures in place. The Western Pacific is the Inner German Border of the twenty-first century, and the Navy should receive a bigger share accordingly, to build the ships and aircraft and weapons systems to take on the Chinese in what they consider their home waters, while keeping the peace and upholding good order at sea worldwide and preventing the Chinese and Russians from establishing their own rules for the global economic system. A good first step would be reversing the decision to exempt the Army and defense agencies from paying their fair share of the strategic nuclear fund, currently burdening only the Navy, Marines, and Air Force.

Finally, the Navy that will carry out the strategy needs to be a *proud and gung-ho* outfit. Going to sea in a warship to protect or to fight is demanding and dangerous, but it's also exhilarating and rewarding. With the proper strategy, force goals, and affordability measures in place, Sailors will rise to the challenge — as they did in the '80s — and accomplish miracles. They will roam the seven seas and visit exotic ports to maintain and expand the US-led global naval and economic system, helping the economy back home as well. They will steam in harm's way to defuse a crisis or deter a hostile power or terror gang from attacking a partner, ally, or US citizen in trouble. And they will sail into combat knowing that they will prevail, given the quality of their strategy, their platforms, their sensors, and their weapons, and the backing of the American people. To reach the American public and its representatives, we saw collaboration with Hollywood, National Geographic, and other American cultural icons as essential (not just some sort of PR frill) to inform and inspire all Americans — as well as our own Sailors — by what my "sea daddy" Admiral Bud Zumwalt was fond of calling the "fun and zest" of going to sea and serving one's country and the free world.

With this superb and provocative book, Dr. Jerry Hendrix has made a major contribution to the contemporary naval policy debate in this country. I hope that readers will be as energized by its concepts, persuaded by its arguments, and motivated by its recommendations as I am.

Preface

Our understanding of the high seas and the laws, rules, and norms that govern them are under pressure. China, Russia, Iran, and others are attempting to reinterpret key sections of the United Nations Law of the Sea and reject legal rulings of international courts in order to set aside long-accepted principles such as free navigation and a free sea (*mare liberum*). They take these actions because they fear a free sea as a medium of transmission of ideas such as freedom and individual liberty. As authoritarian societies, they seek instead to extend sovereign control over the sea as a means of fire-walling themselves and their societies against ideas that might undermine their rule. They have chosen this time to make this attempt because they perceive that the commitment of the United States — the architect of the current liberal international order and the only nation among the three current "great powers" that can claim to have once been a seapower — to a free sea is waning. They believe this because the United States has allowed its naval battle-force, its primary enforcement mechanism of maritime norms and rules, to fall to its lowest levels since World War I, reaching a nadir of 271 ships in 2015, down from 592 ships in 1989.[1]

This massive retraction in fleet size necessarily resulted in a significant reduction in forward deployed ships, which had provided naval presence in every maritime region where the concept of a free sea had been challenged. Naval presence is a crucial military mission that is unique in the modern world to the US Navy. For the past seventy years, in war and peace, the United States Navy has been on station continuously around the globe, preserving and protecting the vital national interests of the United States. In peacetime, while the Army remains in garrison and the Air Force trains and exercises, the Navy remains persistently at sea, reminding every nation by its very presence that the seas remain free and that while some might wish for war to achieve political aims, today is not the day to begin one. Again, if China and Russia, separated from the United States by two great oceans, are indeed the rising great powers that present the most imminent strategic challenge to the United States, then those challenges are best met by the country's Navy.

US Navy ships are positioned to meet these challenges because they possess a distinct advantage over their land contemporaries in that they can exert influence ashore without having to be physically tied to the land. Not only does sovereignty move with each commissioned ship but also through the effects of its sensors and weapons; it can project influence simply by being present offshore. Think of this influence as an incandescent lamp moving about upon the sea. As it approaches an object, its influence can be understood as the degree to which it is sensors and weapons fully "illuminate" or make clear the local strategic environment while demonstrating US interests at the local area. Different types of ships project differing levels of luminance or presence. Aircraft carriers, with embarked air wings that can launch both sensors and weapons to medium ranges, are the brightest lights in the world, while submarines, built to operate with great secrecy, are purposefully designed to cast little to no light whatsoever.

Ships moving toward an area of interest cast a "bow wave" of influence ahead of them as they approach, projecting their capabilities and potential for action well prior to their arrival, yet ships departing an area also leave behind influence in the good will and stability they fostered but also because of the implicit promise that they can, and will, return. Of course, the nation and the Navy cannot afford to have a fleet of 300 aircraft carriers, nor could its influence missions be accomplished by 400 rowboats. Therefore, the size and composition of a navy is a crucial component of a nation's conception of its own power and its strategy for confronting challengers. A nation's or navy's "influence" can be judged by how well it deploys its fleet across its global portfolio of interests. Too small a navy or poor management of its deployment schedule can cause its influence to wane in key regions, thus requiring a period of overinvestment in the region to reestablish the previous accepted level of influence. This is the position in which the US Navy finds itself today.[2]

Such influence can only be regained through the rapid upbuilding of the present Navy, which hovers today around 295 ships, to a 450-ship Navy by 2040. But there must be an understanding that the Navy of the future should not look exactly like the Navy of today, only bigger. It should be a Navy that seeks to retain the best and most lethal aspects of the current fleet, sheds legacy platforms with declining effectiveness, and integrates new technologies and concepts of operations rapidly into the fleet while keeping an eye toward strategic realities and budgetary requirements. And it must be a fleet that balances the twin mission requirements imposed by circumstances upon the US Navy: the need to be technologically advanced enough to decisively win the nation's wars while also remaining large enough to maintain the peace through day-to-day presence. No single ship, no matter how technologically advanced it might be, can be in all places at all times, maintaining all interests. This fact harkens back to the truism that quantity has a quality all its own.

Investment in upbuilding the Navy is required if the current global trading system is to survive. It is a trading system that is built upon the idea that goods can be bought in mass at a low price where they were plentiful, transported more cheaply by sea than by any other mode of transportation, and then sold at a higher price where they were scarce. It is this system that over the past seventy years has resulted in the greatest increase in the global standard of living in human history.[3] However, it is also a system that rests on the assumption that goods can flow without fear of interruption. This assumption is being challenged by the current rise in maritime interdiction incidents as well as Russian and Chinese threats to close the seas (*mare clausum*) near their shores.

The ancient Chinese philosopher Sun Tzu, whose words still provide insight into the Asian way of war, wrote, "The skillful leader subdues the enemy's troops without even fighting."[4] Today China and Russia seek to weaken and then defeat the West by slowly transforming its rules to their benefit. They calculate that they can close the seas near their shores, pushing Western naval powers far from land even as they coerce commercial shipping companies to transit their new territorial waters under new rules, perhaps even paying a nominal piloting transit fee to avoid the expenses associated with longer shipping routes. By pursuing this strategy, they believe that they can successfully "boil the frog," slowly bringing their neighbors and then the world under a new rubric of rules that benefit them as authoritarian powers.

This can be avoided. China and Russia depend upon the complacence of the United States. All that is required is for America to awaken to the present danger, to reorient its strategic focus toward the sea as it once did early in its history, and to upbuild its Navy quickly. If it does not, however, the prognosis is grim. In the absence of American leadership and American sea power, the seas will close, limiting commercial trade and the exchange of free ideas. Spheres of influence will expand and as they do, international stability will decrease. The sinusoidal patterns of history, which have been put off for seven decades, suggest that under these conditions a global maritime conflagration will commence, and oceans will, once again, become battlefields.

A Difference in Perspectives: Land versus Sea

I n the late nineteenth and early twentieth centuries an important debate emerged about the relative importance of sea power versus land power in the ongoing multipolar great power competition. The Napoleonic wars had triggered the creative energies of men such as Antoine-Henri Jomini and Carl von Clausewitz to pen their classics, *The Art of War* and *On War* respectively. These men viewed the competition between nations as one that focused on the capturing and controlling of land and resources. In their minds, capturing was done by armies, and exerting control was necessary to seize the wealth of the land, whether in the form of pastoral grazing land or cropland at a time when agricultural economies were barely beyond subsistence farms, or, later, precious resources such as gold or silver. Despite the fact that "trade" as a wealth multiplier already existed along the trans-Asia Silk Road and on the sea route around Africa and into the Indian Ocean, Europe and Asian leaders still viewed land, especially that which existed along or near their borders, as the primary source of wealth and power. Perhaps this stemmed from the fact that most leaders were monarchical or imperial in character, systems that historically equated land to wealth and wealth to power. This view continued into the twentieth century when Halford Mackinder, a British geographer considered to be the father of geostrategy, promoted in 1904 a revisionist continental great power theory centered around the Ukraine region as the breadbasket "heartland" of the Eurasian continent. Mackinder believed that whoever controlled the heartland controlled the "world island" of Eurasia, and that whoever controlled the world island controlled the world.[5] From this perspective the competition between nations was, of necessity, over land, and the strategy that guided that competition was land centric.

However, the century that followed Napoleon's final exile to St. Helena in 1815 was dominated by a growing appreciation of the sea as both a source of, and a route to, national power. Historians and strategists began to look to Athens, Carthage, Venice, Portugal, Spain, and ultimately Great Britain, and found different variations of a common theme of national greatness built upon a foundation of seaborne commerce protected not by armies but by ships, squadrons, fleets, and navies. Some of these examples, such as the Spanish and, to a lesser extent, the Portuguese, represented extractive models, where the wealth of a newly "discovered" country was simply removed by its new colonial overlords and transferred to the storehouses and vaults of Madrid and Lisbon. The others, guided by the eighteenth-century Scottish economist Adam Smith's vision of a global free market largely unrestrained by protective tariffs and later by British economist David Ricardo's early nineteenth-century writings on comparative advantage, created trading systems that built first their wealth and then their power through the establishment and growth of seaborne trading empires that bought goods at a low price where they were plentiful and sold them for a higher price where they were scarce.[6] These concepts are foundational to understanding the modern Western trading system. However, maritime trade along well-defined routes determined by winds and geography was vulnerable to piracy. It was this vulnerability to interdiction that stimulated the establishment and growth of organized navies to provide protection of chokepoints such as the straits of Gibraltar and Hormuz, where geographic features forced shipping traffic into vulnerable locales. In times of war navies evolved to either maintain their own commercial and logistical sea lines of communications or, alternatively, to interdict those of their enemies.

As the nineteenth century gave way to the twentieth, strategic theorists Alfred T. Mahan and Julian Corbett emerged to offer their distillations of the lessons of sea power from the historical record. Mahan, a US Navy captain, emphasized the possession of a navy of sufficient size, composition, and concentration in order to control the sea and win battles decisively at the moment of a nation's choosing as the critical characteristic of a seapower nation, and thus a great power.[7] Corbett, a lawyer by training, viewed sea power as a support to land power. Based upon his reading of the United Kingdom's experience in European wars, he believed that victory could not be won at sea but could be lost there if maritime supply lines supporting land armies were cut. Hence in Corbett's world, navies existed to protect one's own sea lines of communication, to interdict those of the enemy, and to give direct fires support to land forces ashore.[8] In the end, it was the competition of the ideas of Mahan and Corbett that drove the major strategic dialogues of the twentieth century as the sea and the maritime economy began to knit the nations of the world together in a manner not seen since the fall of Rome destroyed the interconnected trade and security systems of the Mediterranean that it had fostered.[9] From the dawn of the twentieth century until the end of World War II, the international strategic dialogue heavily tilted toward maritime topics as Europe's great powers sought to manage and protect their overseas colonies and the trade routes that connected them. These conversations early in the century focused on raw resources, manufacturing hubs, consumer markets, the ships to connect them all, and the naval fleets required to protect the sea lines of communications.

In the seven decades that have followed the European and Pacific battles of World War II, the world has witnessed a Cold War, waves of terrorism, a great recession, and most recently a global pandemic, but through it all oceanic trade has led to a steadily rising quality of human life on the planet due to greater economic output, the rise and distribution of wealth, and the advancement of science and technology. These factors have led broadly to more productivity, higher literacy, longer lives, and greater ease. Along the way the dominance of the US Navy, which emerged from World War II with a fleet of some 6,000 ships, has been constant and, as a result, the world's five oceans and numerous seas became historically peaceful. Aside from a twenty-year period that began in the 1970s and ended in 1989 when the Soviet navy sought to challenge the American naval fleet, the sea has not witnessed serious competition and has remained peaceful largely due to the overwhelming size and capabilities of the Navy. Until very recently, investments in global navies and naval power have continuously declined even as the numbers of commercial ships have expanded. *Pax Oceania* became the assumed global condition.

Unfortunately, its counterpart, *Pax Terra*, has not been experienced. Despite the lack of a regional or "world war" since 1945, a condition that has largely been attributed to the introduction of nuclear weapons and their general deterrent effect, small transnational wars, civil wars, and counterterrorism campaigns have nonetheless remained a near-continuous factor of modern life. The Middle East has been a cauldron of Israeli versus Islamic strife or Shia versus Sunni conflict for most of the past seventy years. Africa in the postcolonial age has been rife with struggles as historic ethnic boundaries have replaced the convenient borders established by nineteenth-century imperial European overlords. Central and South America have experienced domestic political upheaval and significant racial and social movements. Europe, although largely peaceful, has recently experienced "gray zone" warfare in Georgia, the Crimea, and Russia's outright occupation of other areas of Ukraine through the use of special forces wearing no identifiable military insignia, commonly referred to as "little green men," to stir up trouble and dissent. The three Baltic states of Latvia, Lithuania, and Estonia, as well as Poland, all formerly heavily terrorized and subjugated by Russia, have sought to strengthen their militaries and to maintain a constant eastward guard. Asia, especially southeast Asia, experienced great tribulation following their escape from colonial control. The Korean peninsula continues to exist in a state of war controlled only by an unsteady armistice. China, the largest and most ancient of Asian powers, claims only peaceful intent with its continental neighbors despite aggressive language and actions toward many of them, which is good because its forces have not done well or have experienced mixed results in combat against Vietnam (1979), Russia (1969), and India (1962). China recently skirmished again with India high in the Himalayan range, apparently in hand-to-hand combat, in a situation that threatened to escalate.[10] China's one modern moment of ground combat success was its introduction of ground forces on the side of the north during the Korean War in 1950.[11] Finally, North American relations have largely been peaceful and even cooperative but it must be acknowledged that both the United States and, to a much lesser extent, Canada have found numerous reasons to conduct overseas military campaigns, some lasting decades.

Korea and the US Army's operations there began a new era of American overseas military action. Whereas in the past the US Army had mobilized and expanded to meet the demands of World War I and World War II, it had also contracted severely immediately following both wars.[12] This was in keeping with the Founding Fathers' constitutional injunction against large standing armies and George Washington's advice that the nation not "forego the advantages of so peculiar a situation? Why quit our own to stand upon foreign ground?"[13] The crossing of the artificial boundary between North and South Korea at the 38th parallel along with other rising concerns about broken Soviet promises in Eastern Europe convinced the Truman administration to reverse course and begin building the Army up again. While Eisenhower did attempt to tamp down national expansionist tendencies during his eight years in office, Kennedy's inaugural clarion call to "pay any price, bear any burden, meet any hardship, support any friend, oppose any foe" led to attempted incursions into Cuba as well as growing commitments in Vietnam, reversing the nation's long-held reluctance to intervene overseas with land forces and the subsequent permanent establishment of a large standing army. Such a large peacetime army has continued to present such a temptation that even the most dovish of statesmen came to cast a covetous glance at soldiers in garrison.[14] Such glances led to foreign policies from both the left and right political spectrum that committed troops to land wars throughout Asia over the past seventy years.

Oddly enough, the causes of most of those land wars were not the historical hegemonic acquisitions of territory and resources for the state but rather the unification of historical nations that had been rendered asunder by colonialism, the breaking up of modern states in order to restore historical boundaries of smaller states, the pursuit of ancient ethnic grievances, or attempts to settle ideological or religious differences by force. Only Russia's seizure of territory in Georgia and Ukraine and China's annexation of Tibet stand out in the post–World War II era for their rejection of Westphalian norms with respect to the territorial sanctity of the nation-state. The effect of these numerous ongoing campaigns as well as their frequency, if not simultaneity, resulted in a heightened awareness in the public mind of land combat operations.

In the aftermath of the counterinsurgency war in Vietnam the US Army, despite its long history of "small war/counterinsurgency" campaigns, was anxious to return to planning for large wars and a European focus. As such, in the late 1970s, under the leadership of US Army Generals William DePuy and Donn Starry, it conceived "AirLand Battle" to address the Soviet overmatch in terms of personnel in a prospective European war. While the rapid disintegration of the Soviet Union in 1989 took away the immediacy of purpose for AirLand Battle, the synergies it birthed provided the foundations for an overwhelming victory in Operation Desert Storm, which was waged against a modern and large (fifth largest in the world) Iraqi military force that had just emerged from a decade of hard desert combat with its neighbor Iran.[15] The success of this campaign — in particular the combined effects of stealth aircraft, precision strike weapons, and massed maneuvering formations of armored vehicles — suggested to some that a new "revolution in military affairs" was occurring that would allow modern militaries to outdistance traditional strategies, concepts of operations, and tactics.

However, a decade later the strategic shock and tragedy of September 11, 2001, and the subsequent entry of US forces into combat, first in Afghanistan and then later on Iraq, threw the strategic community into an era of confusion and introspection. Initial entry into counterinsurgency operations led to a reexamination of the "eating soup with a knife" tactics of T.E. Lawrence, as well as multiple iterations of leadership changes before a "clear-hold-build" strategy could be implemented.[16] Later, the integration of unmanned air and ground platforms as well as the widespread use of human enhancement techniques led to new concepts of operations focused around human-machine teaming with the potential of approaching the ideal of a future environment that maximizes lethal effects upon the enemy while minimizing risks to one's own forces.[17]

Most recently, combat futurists have begun to promote a combination of lethal weapons, cyber capabilities, and space-based sensors that could create a cheaper, yet effective, military system of systems.[18] All of these concepts present promising new insights and even strategies for the conduct of war. Most are land centric. Few of these initiatives, aside from the defunct AirSea Battle concept, which was terminated due to Army objections that it did not sufficiently include the land component of the Joint Force, considered the naval environment, which is odd.[19]

The Sea as the Source
of the Wealth of Nations

Anyone looking at one of the iconic photos from the Apollo moon landing era taken by astronauts in lunar orbit looking back upon the earth will be struck by the fact that the planet's appearance is dominated by its blue oceans. The oceans make up 72% of the planet's surface and while land-centric strategists would be correct to point out that nearly 100% of humans live on land, a closer look would reveal that some 2.4 billion people, or 40% of the population, live within 60 miles of an oceanic coastline.[20] What is more, with a global economy of well over $86 trillion, some 80% of global trade by volume and 70% by value travels over the sea and into coastal port facilities.[21] This last point did not occur overnight, and it also did not evolve accidentally. This level of trade occurs because the sea is a safe environment in which billions of dollars in investments in ships and surface and subsurface infrastructure can be protected and preserved.

Following World War II, the global economy was in ruins. What economic activity there was had been focused on supporting the war effort and the shipping hulls that existed were preponderantly operated as logistical support for military operations on the European continent against Nazi Germany or across the Pacific fighting imperial Japan. However, as the wars raged, leaders were determined that a new global environment would emerge after the strife that would not be based upon colonial exchanges but rather upon free trade among all nations. At a conference held near Bretton Woods, New Hampshire, the United States proposed a postwar "dynamic world economy in which the peoples of every nation will be able to realize their potentialities in peace and enjoy increasingly the fruits of material progress of an earth infinitely blessed with natural resources."[22] As a result of the conference, the World Bank and the International Monetary Fund were founded and the beginning of what would become the General Agreement on Tariffs and Trade (GATT) was outlined. The GATT in particular opened economies that had once been closed and allowed for the increased movement of goods and investment by large corporations into nations and regions with large amounts of raw energy and mineral resources, driving up requirements for a larger, more efficient, and more specialized shipping system.[23]

Following the war, the global economy made use of the existing war-built logistics fleet for most of its shipping needs, but as time passed, demand for steel and aluminum as well as a growing preference for imported oil over domestic coal drove the growth and specialization of the bulk merchant fleet. The case of oil tankers is illustrative. In 1900 the largest oil tanker displaced around 12,500 tons and the size of this class of ship grew gradually over the decades that followed until 1944, when the largest tanker displaced nearly 24,000 tons. However, the growing demand for oil over coal as a power generation and heating source resulted in the building of the 123,000-ton tanker *Universe Apollo* in 1959, the 327,000-ton *Universe Ireland* in 1968, and the 556,000-ton *Seawise Giant* in 1980.[24] The greater cost efficiency of ever larger hulls resulted in the loading, transporting, and delivery of raw resources to manufacturers at a vastly lower cost than what could be achieved via overland transportation methods. Specialization of ship construction soon led to the creation of ships carrying pure ore, grain, sugar, liquified natural gas, and chemicals.

Improvements in shipping did not stop with ships designed to carry raw resources. During the 1960s transport companies began to develop a standard "container" that finished goods could be packed into and transported by truck from factories to train stations, where the containers could be transferred to rail flatbeds and thence transported to port terminals, where the containers could be loaded onto ships designed to carry them in ever increasing numbers. The first postwar container ship carried 58 metal containers from Newark, New Jersey, to Houston,

Texas, in 1956. Today the largest container ships can carry up to 14,500 containers, each 20 feet long by 8 feet wide by 8.5 feet tall.[25] Again, with engineering plant improvements, smoother new hull designs, and advancements in navigation, placing more goods on ever larger ships represented the most cost efficient method of moving products from geographic regions where they are cheap to produce to other regions where they are rare and bring a higher selling price.[26]

It must be noted that growth in trade was generated not just in the increased size of ships, but also in the number of larger ships built as the growing global economy created demand for more goods at lower prices, hence a requirement for ships to transport those goods. In 1950 the overall tonnage of the global merchant fleet was around 85 million tons, but by 2005 it had risen to 653 million tons.[27]

This growth in maritime trade was driven by three factors. First was price. As the size of merchant ships increased, along with the amount of cargo they could carry, the cost to transport a set quantity of goods came down. For example, during the 1960s the amount paid by a Middle East–based oil company to cover its insurance and freight costs to transport a barrel of oil from the region to Europe amounted to 30% of the final price, but by the 2005, as oil tankers grew in size, that cost had fallen to around 5%.[28]

Another characteristic that drove the expansion of maritime trade was reliability. With new "just-in-time" supply models, many wholesalers came to regard the container ships carrying their goods as their warehouses, thus avoiding local storage costs. Shipping companies with well-mapped routes measured by global positioning systems that included built-in adjustments for foul weather became very confident of the schedules assigned to their ships.

The final factor driving maritime trade growth was security. Loss of cargo or damage at sea can be insured against, but over the past seven decades, confidence in the security of merchandise sent by sea, even the highest in value, fragile, or rare, has grown stronger with the passage of time. Today's participants in the global economy place their trust in seaborne trade because they believe that it is the cheapest, most reliable in terms of schedule, and safest method of transportation. However, from a maritime perspective, wealth and the sea are no longer restricted to travel upon its surface. In the past few decades, the bottom of the sea has become yet another source of wealth as well as a transfer plain for data information.

The British led the way in understanding the importance of having the most accurate information through the quickest method. Long dependent upon swift packet ships to bring news from far abroad, the British laid the first underwater cable across the English Channel in 1850. Within a few years, cables were laid to link Great Britain with Ireland, Belgium, the Netherlands, and then Denmark. British bankers and investors were wise to the advantages of learning new information swiftly and then adjusting their positions within stock markets to maximize earnings. Within fifty years Britain had knitted together a global empire, upon which the sun never set, with underwater cables. As the twentieth century progressed, these cables grew larger and more plentiful, allowing increased amounts of information to travel at higher data rates, to the point that voice communications between the continents became possible. During the 1980s, the first undersea fiber-optic cables were laid, and data information rates climbed exponentially.[29] Today the information economy represents around 16% of the global GDP and is worth approximately $11.5 trillion.[30] Within that sector of the economy, about 95% of internet data and voice information that travels between continents moves along undersea cables, rendering the cables themselves both incredibly valuable and vulnerable to tampering or destruction.

Of course, the seafloor's full value is not limited to information that travels across it but can also be found in the raw resources that lie under and upon it. The first oil wells built over water occurred in freshwater lakes in Ohio during the oil boom of the 1890s. The first wells drilled from oceanic platforms occurred just off the beaches of Santa Barbara in 1896. For most of the twentieth century, oil wells proliferated just off US shores where the depth of the water was largely less than ten meters. It was not until the latter half of the century, when designs for stable floating drilling rigs matured, that offshore drilling came into its own.[31]

Today, oil and natural gas extraction occurs in depths of nearly 6,000 feet, with major fields off of California and the Gulf of Mexico, the North Sea, the Orinoco Belt field off of Venezuela, the Arabian Gulf, and most recently in Arctic Ocean waters off of Alaska, Canada, and Russia.[32] These energy "fields" cover large areas of the oceans, with individual seafloor wellheads being connected by a vast array of pipelines that generally neck down into a single underwater artery that pumps the product ashore to a central refinery.[33] Oil and natural gas production currently contribute over $3 trillion in income to the global economy each year, a number that is expected to rise in the future.[34] With much of the world's untapped reserves being located underwater, the resources that lie under the seafloor, such as those in the Arctic Ocean, which is beginning to open to energy exploration due to the effects of climate change, will be a major source of both wealth and competition in the decades to come.

In addition to oil and natural gas extraction, which for some has become "normal" despite the forbidding conditions, many companies whose names are more associated with mining on land, such as the South African De Beers Group, have begun to pursue resources on the seafloor. In 2018 De Beers was able to gather 1.4 million carats of diamonds off the coast of Namibia and in 2019 they increased their efforts. Another company, Nautilus Minerals, mines a series of underwater hot springs off Papua New Guinea to gather gold, silver, and platinum traces that are welling up from deep inside the earth. New methods of extraction as well as classes of ships have been built to lift copper, manganese, nickel, cobalt nodules, certain rare earth elements, and other resources from the ocean floor, often miles below the surface.

With 85% of the ocean floor unmapped, scientists believe that $150 trillion in gold could be extracted in the years ahead.[35] As methods, technologies, and vehicles mature, nations will rush to develop the resources contained within their extended economic zones, roughly defined as an area extending out to sea 200 nautical miles from a nation's territorial sea baseline, but in reality the initiative to mine raw resources from the ocean floor is already outpacing the current rules and laws as nations and multinational corporations race to reach resources simply lying on the seafloor in international waters.[36]

As an aside, it's important to note that the United States has not ratified the United Nations Law of the Sea treaty largely because of Part XI of the treaty, which dealt with the economic development of the seabed and which the United States feels is injurious to its free market interests.[37] However, the United States continues to support the other aspects of the treaty despite not being a signatory.

What does this all add up to? The causal link between the sea and improvements in the human condition is so obvious as to beg its assertion: Oceanic security is a prerequisite for economic growth, and economic growth leads to improvements in the human condition. Since the end of the last great oceanic war in 1945, the percentage of the world population living in extreme poverty has fallen from 72% to 9.6%.[38] Over that same period, world literacy has risen from 50% to nearly 88%.[39] Over that same era, global life expectancy, perhaps the best metric of humanity's success, has risen from 47 years in 1950 to 74 years today.[40]

What must also be said is that the unfettered growth of economic activity under and upon the sea are undergirded by two assumptions: that the seas are free, and that they are safe to operate upon. No individual or corporate entity would spend the millions or billions of dollars to build an oil drilling platform or lay a cable if they thought there was a high likelihood that it would be attacked, stolen, or destroyed. *The world economy grows because the sea as a global common is safe and free to all.*

The Creation
of a Free and Safe Sea

The concept of a free sea, by which we mean that individuals, companies, and nations have the right to cross international waters without fear of being stopped, searched, or delayed in any way so long as they operate within internationally accepted laws and norms, has been accepted part of the Western canon for four centuries. International waters are defined as the sea and oceanic commons that lie beyond a nation's legally claimed territorial seas. These regions of maritime sovereign control were recognized under law and allow nation-states to exercise both defensive and legal controls along their oceanic borders, just as they are legally able to control the territorial boundaries along a mountain range or river dividing two countries. While the concept of a free sea appears to be common sense today, it was not always so, and there is more than a hint that the concept is under fresh legal challenge today.

The philosophical principle of a free sea emerged during the Roman era, but it was through the effort of the Dutch legal theorist Hugo Grotius that it gained acceptance within the modern enlightenment construct of natural law.[41] Grotius took the initiative to lay out his arguments as a rebuttal to Portugal's *mare clausum* policy and their claim of a monopoly over all trade with the East Indies. Portugal based its legal claims upon the Treaty of Tordesillas, signed by Portugal and Spain in 1494, which divided the world between the two great Catholic maritime powers along 46° west longitude. This line was modified by the Inter Caetera papal bull of Pope Alexander VI in 1493, which moved the line slightly eastward to the 38° west longitude.

These rulings prohibited new powers, to include the Protestant Dutch, from pursuing trade on the high seas and thus led to rising tensions between nations. In 1603 the friction emerged into the open when a ship owned by the Dutch East India Company, one of the world's first companies backed by publicly traded stocks, captured the Portuguese ship *Santa Catarina* in the nominally Portuguese controlled Strait of Singapore. Its cargo was valued at three million Dutch guilders, which was comparable to the annual tax receipts of a small nation at the time. A case was brought as to the legality of the capture, given Portugal's claim of a monopoly over trade in the Far East, and Grotius was hired by the Dutch East India

Company to defend its actions. In a remarkable legal sleight of hand, Grotius advanced an argument that lost his client its prize, the cargo of the *Santa Catarina*, while gaining that same client, and all maritime powers, guaranteed access to a free sea.

Grotius grounded his line of legal reasoning in natural law, a philosophy that argued that all laws originate from God and can be recognized in nature. He then advanced the idea that although peoples are allowed by nature to divide into nations, those same populations, like nature's wind, are also allowed to blow unencumbered from place to place and trade with each other, and that no prince may block an individual arriving at his border from trading with his people. Building upon this principle of innocent passage over land for the purpose of trade, he then advanced that a similar right must be allowed over the seas, which lacked in nature even the natural boundaries of mountains or rivers, sharing more characteristics with air and thus commonly shared by all.

Grotius then went on to argue that the randomly changing state of the seas revealed that they cannot be possessed, and that the bottomless nature of their depths and the animals that lived there suggested that it could not be "used" or harnessed and hence become property. Thus, the sea could only be considered a common area and neither Portugal nor any other nation could "own" or exercise imperium in any part of the sea. By extension, without a right of national ownership or control, governance reverted to God, and hence there could be no right to refuse a free crossing or navigation of a portion of the globe commonly accessed by all men under God.

In the end, Grotius's argument allowed the value of the cargo of the *Santa Catarina* to return to Portugal, but his legal argument established the concepts of *mare liberum*, free trade and free navigation, all emerging from natural law, as foundational principles of the Enlightenment.[42]

Grotius did not have the last word on the topic. Later theorists, arguing on behalf of national interests, continued to advance the claim that nations could gain territorial sovereignty over vast portions of the oceans so long as the case could be made that maritime territoriality was attached to legally recognized land territoriality claims and that the nation in question was capable of exercising and maintaining control over both of them.[43] Later, a second Dutch jurist, Cornelis van Bijnkershoek, provided a more moderate approach to territorial sea assertions by suggesting that sea control by ships was necessarily transitory in nature and as such territorial seas should be limited to the control that could be exercised over them by permanent land-based weapons. "I hold that the territorial dominion ends where the power of weapons terminates."[44] This resulted in what came to be known as the "cannon shot rule" whereby states could claim seas outward to the maximum distance of a shore-based cannon. Over time this resulted in the recognition of one-league or three-nautical-mile territorial seas, a norm that remained in place until the 1982 United Nations Convention on the

Law of the Sea (UNCLOS) established twelve nautical miles as the definition of a nation's territorial sea.[45]

However, the establishment of the principle of a free sea and the assumption or expectation of a free sea are two separate concepts altogether. The first was established in the century that followed the publishing of Grotius' argument. The latter has taken nearly three centuries to solidify.

Maritime conflict was the norm during the seventeenth century as rival commercial powers in Europe struggled for supremacy. After surging to the fore during the fifteenth and sixteenth centuries to pioneer routes down the west African coast around the Cape of Good Hope into the Indian Ocean, the Portuguese empire had receded in the seventeenth century as Spain moved aggressively outward in terms of both exploration and colonization.[46] Spain's moment of dominance was brief however, as the Dutch, the English, and the French each moved to gain access to resources, charter colonies, and build navies to protect trade. Competition was fierce as 226 naval battles were fought between the rival seapowers, with most of them waged between the English and the Dutch as they struggled to gain sea control over the waters between the British Islands and the Low Countries.[47] England emerged triumphant in this competition but by century's end France had risen to nearly equal the island nation in terms of naval gross tonnage at 196,000 and 195,000 tons of ships under sail respectively. The Netherlands were respectable at 113,000 tons but Spain had effectively withdrawn from the naval great power competition at that point, floating only 20,000 tons of warships.[48]

The eighteenth century was marginally better in terms of the number of naval battles, with only 138 instances of naval strife; however, the size and intensity of those battles increased as the technologies and concepts of operations for naval warfare matured. Ships went from single-deck sloops, brigs, and frigates to multi-deck line of battleships, and, just as the name implies, battles went from single ship-on-ship melees to organized lines of battle as squadrons, divisions, and fleets organized themselves and then closed to wreck devastation. For example, the Glorious First of June battle in 1794 between Great Britain and France, each fielding 25 ships of the line, came to epitomize naval battle as the Royal Navy captured 6 French ships and sank another while killing or wounding some 4,000 French sailors and capturing another 3,000.[49] Increases in the severity of war were accompanied by the growing size of European navies. In 1770 the fleets of Europe combined to displace 750,000 tons but twenty years later this metric had more than doubled to 1,700,000 tons as competition between Great Britain and France accelerated.[50]

Great Britain's final defeat of imperial France did not occur until 1815. This is important to the discussion, because more than half (73) of the entire nineteenth century's 140 naval battles occurred within its first fifteen years. The century's third decade witnessed another 32 battles associated with the Greek war for independence. The rest of the century witnessed few naval battles even when the bloody American civil war is considered, and the battles themselves, with the exception of Japan's naval battles with China in 1894 and the United States' two battles with Spain in 1898, were relatively small in terms of the numbers of ships involved. Due to Great Britain's global leadership, most of the nineteenth century was relatively quiet and saw a period of industrial expansion accompanied by increased maritime trade.

Britain's dominant role in the world throughout the nineteenth century can be traced to several factors. The first was the impact of the industrial revolution, which allowed the island nation to lead the world with a century of average 2% GDP growth.[51] The industrial revolution also allowed Britain to take the lead in the development of steam-powered ships clad first in iron and then, taking full advantage of the industrial revolution at home, in steel, carrying ever larger rifled naval guns capable of shooting accurately at greater distances. These new ships could "sail" against the wind, changing fleet tactics and methods of operations forever.

The second change, which occurred at the strategic level, involved pursuing a strategy that attempted to avoid entanglements in continental politics and its numerous land wars, restricting Great Britain to the seas where it recognized that the island nation's strengths lay. In short, Great Britain sought to thoroughly control or, perhaps more delicately, "administer" the seas without becoming entangled on land. Looking back over the previous two centuries, the British noted that it was during those periods of time *when it did not enjoy overwhelming supe-riority at sea that it was most likely to be challenged.* Given the nature of its far-flung global empire, built upon the principles of a free sea and free trade and upon which "the sun never set," the government realized that it could not afford instability at sea. As such, the British government set about establishing global norms for accepted practices on the high seas codified into admiralty law and then took the next step of establishing a naval architecture policy that mandated that the Royal Navy be equal to or greater than (in terms of tonnage) the next two navies combined in order to create a mixed constabulary/war fighting force that could enforce those norms and laws. This resulted in what became known as England's "Two Power Standard."[52]

However, broad support for the concept of "sea power" as a component of the imperial territorial and economic competition soon led other great powers of Europe, including Germany, Italy, and Russia, as well as China and Japan in Asia, to build fleets that would allow them to gain and maintain control of overseas colonies or other forms of territorial possessions as sources of wealth and interna-tional power. Soon the nations of Europe and Asia (Japan in particular) engaged in diplomatic negotiations to cre-ate larger "spheres of influence" in Africa, the Middle East, and China.[53] These spheres essentially codified colonialism as the accepted global governing norm, but it also caused long-standing competitions between European nations to be distributed worldwide, with the sea becoming the pre-ponderant medium of transfer. Soon the world witnessed increased international instability that spread to the seas despite Britain's "Two Power Standard" as smaller powers retained the potential to combine capabilities and overmatch Britain, and the nature of the global competition required Great Britain to maintain naval presence everywhere while lesser powers such as Germany could focus their power in a limited area of operations such as in the North and Baltic Seas, and thus negating British leadership. These factors combined to create a naval arms race at the dawn of the twentieth century.

At its heart, this race arose from the increasing complexity of international relations between ever greater powers, especially during a diplomatic era wherein foreign policies were not the expression of national principles but rather personal expressions of sovereigns or other large personalities in strong governmental positions. In this era foreign policies were often characterized by what or who they were "against" rather than who or what they were for. For instance, it had been Great Britain's policy to avoid participation in Europe's land wars while acting as an offshore balancer for Europe, shifting its support from one continental actor to another across the nineteenth century to ensure that no state could achieve dominance as Napoleon and Louis XIV had done in the past.[54] Britain deftly managed the continent for fifty years after Waterloo, but the creation of Germany as a cohesive central European state in 1871 brought about a strong power in the heart of the continent that immediately began to negotiate new treaties and alliances that threatened Britain's global interests.[55]

While Britain had been satisfied to remain detached and unentangled from Europe as long as a continental power did not interfere with its world empire, Germany, especially under Kaiser Wilhelm II (grandson of Britain's Queen Victoria and jealous of her power and prestige), began to focus on building its own global empire in order to find "a place in the sun" for the new state. Germany's diplomatic maneuvering, which included the formation of the Triple Alliance with Austria and Italy in 1882, forced Great Britain into the European alliance it had avoided for so long, the Triple Entente with France and Russia, formalized in 1907. The Entente initially did seem to strengthen the United Kingdom's position on the continent, throwing Germany's foreign policy temporarily into disarray.[56] However, Germany had a new strategy in the offing.

For years Alfred von Tirpitz, a German naval officer and strategist, advocated for a larger and more powerful German navy with the intent of using it as a coercive diplomatic tool against the British overseas empire. Forecasting that "even the greatest seapower would act more accommodatingly toward us if we were able to throw upon the scales of international politics, or if necessary into the scales of conflict, two or three well-schooled squadrons," Tirpitz recommended the construction of several squadrons of eight battleships each for the task.[57] His "risk theory" strategy sought to threaten Britain's overseas possessions to coerce the seapower either to be neutral or perhaps even to ally with Germany's ambitions on the continent.[58] Tirpitz and his enthusiastic sovereign Wilhelm II passed their first naval law in 1898, with supplementary expansion laws coming in 1900, 1906, 1908, and 1912.[59] Due to its stated "Two Power Standard" policy, Great Britain had no choice but to respond in kind with additional naval shipbuilding of its own.

However, as described in Robert Massie's magisterial 1991 book *Dreadnought: Britain, Germany, and the Coming of the Great War,* the Royal Navy went further than competing quantitatively when it introduced the qualitatively superior all-big-gun, steam-turbined battleship HMS *Dreadnought* in 1906, immediately rendering all previous warship designs obsolete. With *Dreadnought's* appearance, Germany found itself chasing Great Britain not only in terms of numbers but also in technology, and Germany was not the only nation in the competition.[60] By 1914 seven countries (Great Britain, Germany, France, Austria-Hungary, Italy, Japan, and the United States) fielded a combined 59 post-Dreadnought battleships with another 41 under construction. In addition, there were 137 pre-Dreadnought battleships still in service across ten nations (the seven Dreadnought nations plus Russia, the Ottoman Empire, and Greece).

In the days leading up to World War I, Britain's attempts to maintain its security through the "Two Power Standard" fell victim to an expanding number of spheres of influence in Africa and Asia. In addition, Germany, whose very origin was premised on the expectation that it would gain hegemony on the continent, rutted about in the Balkans, Baltic, and Dardanelles for opportunities to undermine the British-led Triple Entente. It was Germany's plan to destabilize the balance of power in Europe, creating an opportunity for the expansion of German power. As a cumulative effect of these events, Britain's position of leadership at sea quickly degraded, the global environment destabilized, and the conflagration of Europe began, ultimately resulting in the outright destruction of long-standing empires and a generation of the continent's youth. At sea, 19 naval battles were fought, including the massive Battle of Jutland, which resulted in nearly 9,000 dead and some 180,000 tons of ships sunk in two days.[61]

Following the armistice of November 11, 1918, few naval battles occurred during the interwar era aside from four small battles associated with Spain's civil war. However, the outbreak of World War II more than compensated for the two decades of quiet that proceeded it. From the attack upon the US Navy gunboat *Panay* on the Yangtze river in 1937 to Japan's surrender on the deck of the USS *Missouri* in Tokyo Bay in 1945 there were 97 battles, many of them massive in terms of the number of ships and aircraft participating as well as overall losses. The war heralded an era of huge investment in naval power. The United States grew its fleet from 394 to 6,768 ships during the war. Over the same time the Royal Navy expanded from 137 to 2,340 ships.[62] Among the victorious powers, the Soviet Union was notable in that of the more than 110 ships it built during World War II, half of them were submarines, a sign of things to come for that nation.[63]

Following the war, partially as a cost savings effort but also in recognition of the diminished threat upon the world's oceans, every major power dramatically reduced their fleets. The Royal Navy fell to 453 ships by 1950, 441 by 1960, and 184 ships by 1970, and continued to contract in the decades that followed. The Soviet Navy initially drew down its fleet as it sought savings to recapitalize its war-damaged civilian industries, but by the mid-1950s it began once again to expand its navy in earnest due to the perceived threat from the West, achieving a fleet of some 600 ships of various classes by the 1970s.[64] For its part, the United States dramatically drew down the size of its fleet following the war, falling from over 6,000 ships at the war's close to just over a thousand ships in 1955. Within the Cold War the US Navy reached its smallest fleet of 521 ships in 1981, just as the Carter administration left office. As part of a political election promise by Ronald Reagan, the Navy began an arduous climb back toward a 600-ship goal under the leadership of Secretary of the Navy John F. Lehman, culminating in 594 ships in 1987.

Pax Mare Americana

Following its World War II victories in Europe and in the Pacific Ocean, the United States occupied a position of strength on a nearly unprecedented global scale. The size of its Navy and the absence of any foreign navy that might challenge it created an environment where the United States could establish and uphold rules and norms across nearly all the world's oceans and seas. The ideas of free navigation and free trade became widely accepted international rules and norms. Sea routes that provided the fastest means of transport for people and cargo became accepted even as wars, battles, and piracy were eliminated. People, products, and perspectives began to compete across the globe, unencumbered by fears of interdiction at sea. While wars continued on land on the Korean Peninsula as well as in Africa and Vietnam, and later in the Middle East, maritime conflict all but disappeared on the world's oceans and sea, and much of the credit for this era of peace can be assigned to the United States Navy.

Throughout the post–World War II era, wherever the Navy operated the seas opened to free trade and the liberal economic order that the United States sought to establish. This success created costs as US-backed norms and rules flowed outward into new geographic regions and, in turn, created more demand for naval presence in new regions previously not patrolled. At first the requests for increased presence were driven by the nation's counter–Soviet Union strategy of containment, which sought to limit the expansion of communism around the world. This strategy led to the regular deployments of ships to the Mediterranean and North Atlantic.

A few years later, the desire to protect the Nationalist Kuomintang government on the island of Taiwan and later proxy wars on the Korean peninsula and in Vietnam increased deployment requirements in the western Pacific. The OPEC oil crisis of the 1970s and rising instability in Iran drove the US Navy to declare national interests in the Middle East and establish a permanent naval presence in the Arabian Gulf. Additionally, Soviet and Cuban support of anticolonial movements in Africa and later famine and local wars of ethnic genocide drove US policy makers to begin to operate ships and expeditionary forces in and around that continent. Today regional combatant commanders (four-star generals or admirals heading up the Indo-Pacific Command, European Command, Central [Middle East] Command, Africa Command, or Southern Command) of the United States have attached "national interests" to no less than nineteen distinct maritime regions and requested persistent presence from the Navy within them in one form or another.[65] Throughout the Cold War the United States Navy outpaced the navy of the Soviet Union in terms of fleet size as well as technological capabilities.[66] The size of the post–World War II naval fleet, averaging well over six hundred ships throughout the Cold War, and the evolution of these requirements allowed the US Navy to evolve a deployment model unlike any it had previously utilized.

Early in the American republic's history, the US Navy's preferred operating method was to keep its ships close to home, "surging" periodically to engage threats as they emerged.[67] This model was also used just prior to World War I. In other eras the Navy maintained small squadrons of ships at forward geographic stations the Mediterranean or Caribbean. This model was used during the early 1800s, when the nation had difficulties with the Barbary states and met them with the Mediterranean squadron under Commodore Edward Preble, and at the end of the nineteenth century, when the Spanish American War broke out and the Asiatic Squadron under Commodore George Dewey proceeded to Manila Bay to take on the Spanish fleet there.[68]

Another model employed, especially when the fleet was small, was the "cruising" model, which sent single ships out to "show the flag" and protect American commercial interest in various regions without explicitly declaring a permanent national interest. In the decade leading up to World War II, the US Navy shifted to an experimentation model, keeping the fleet close to home while conducting multiple fleet exercises each year based upon perceived threats in Europe and Asia. Such naval deployment models were appropriate for eras when the United States was not a superpower or even a great power, which explains why they were abandoned following World War II.

World War II itself forced the Navy to change to a forward-operating model. To deal with the European war, ships were trained and deployed from the United States under the direction of the commander of the US Atlantic Fleet. When they arrived in European waters, they were transferred to the operational control of the commander of US Naval Forces in Europe. Similarly, the US Pacific Fleet Commander, whose headquarters was moved from San Diego to Hawaii just prior to the war, was charged with training and outfitting ships as well as developing an overall fleet strategy. However, when it came to commanding the ships in combat, responsibility fell to a three-star vice admiral who commanded a numbered fleet (Third Fleet or Fifth Fleet).

Immediately following the war, the US Navy returned home to demobilize, deploying small groups of ships on long cruises to show the flag, but the worsening of the relationship with the Soviet Union and the advent of the Cold War led national leaders to return to the wartime practice of training and equipping ships on the United States' East and West Coasts (Second and Third fleets respectively) and then deploying ships forward to fleet "hubs" (Sixth Fleet in the Mediterranean, Seventh Fleet in Japan and the Philippines), where they would operate before returning home again for maintenance and training, thus restarting the cycle.

At the height of the Cold War the US Navy maintained an average of 150 ships at sea at any given moment to support the national interests. As the decades of the Cold War passed, the forward hubs became the permanent home to many of the ships and their crews, with maintenance and training occurring on location. For most of the Cold War in the Mediterranean, the western Pacific, and later the Middle East, US interests, and to a broader degree the interests of the West, to include free navigation and free trade, were upheld with relative ease as a large US naval fleet routinely flowed from home waters to forward fleet hubs and thence onto the various distinct maritime regions on a regular and persistent basis with little to no resistance. In fact, during the seventy-five years that have passed since World War II, only twenty-five incidents roughly described as "naval battles" have occurred, and none of them have been major in terms of tonnage sunk and lives lost. What is more, there was only one naval/oceanic war fought during this era of *Pax Mare Americana*, the British versus Argentina Falklands War. However, this peaceful era appears to be coming to an end.[69]

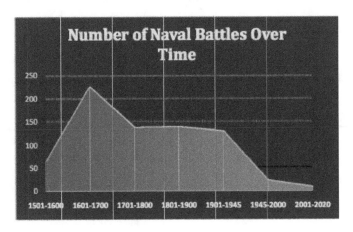

Things Fall Apart; the Center Cannot Hold.[70]

After the Cold War, its participants unsurprisingly sought a "peace dividend" through massive cuts in defense spending. In 1989 the DoD topline budget was $593.34B in 2020 dollars, with the Army receiving $159.6B, the Navy and Marine Corps combined receiving $198.64B, and the Air Force receiving $193.5B or 27%, 33.5%, and 32.6% respectively.[71] The remaining portion of the DoD budget was allotted toward other DoD expenses. Ten years later in 1999 the total DoD budget stood at $387.15B in 2020 dollars, with the Army getting $96.0B, the Navy and Marine Corps combined receiving $122.33B, and the Air Force coming in at $119.92B or 25%, 32%, and 31% respectively, again with an increased remainder going to support other portions of the Department of Defense.[72]

These numbers represented a 35% cut in the defense budget. The US Navy made up a significant contribution to this effort in terms of real infrastructure, decommissioning 100 ships in the three years that followed the Soviet Union's dissolution. The fleet dropped another 100 ships in the next four years, ending 1996 with a 375-ship battleforce. By that point, the downward momentum had accelerated, and the US Navy crashed through the 300-ship barrier in 2003, bottoming, as previously mentioned, at 271 ships in 2015. A point to consider is that when a navy builds ships at a higher rate, say 18 ships per year, 25 to 30 years later that navy will be decommissioning those same ships at the same rate. Unless that navy is building new ships at the same rate, it will shrink in size. The Navy budget of the early 2000s was simply not large enough to finance the construction of new ships to keep up with the retirements of the ships built during the Reagan era naval buildup, and, in fact, a growing portion of its budget was dedicated to providing maintenance for its aging fleet.

During the same period Russia's Navy was laid up, preserving only the most advanced aspects of its shipbuilding industrial base. China's navy during this era focused its fleet first on coastal defense, then on a cross-channel invasion to overwhelm and defeat the Nationalist Kuomintang government on Taiwan, and then on building a defensive buffer, or wall, within the first island chain that extended from Japan to Taiwan to the Philippines and down into Indonesia. During the 1970–1995 era PLAN ships were coastal in their design and intended methods of operations. They were not built to operate in a "blue water" deep-ocean environments, outside of the range of land-based support, and most lacked even the ability to remain at sea for more than a few days without having to return to port to refuel and resupply with food. There was no reason for the Chinese to venture out into those waters. It was a daunting proposition to attempt to compete with Americans.

Throughout the Cold War and even into the first years of the inter–Cold War era, the sheer size and overawing technological capabilities of the American Navy ensured that no foreign power tried to compete with it at sea. It was so dominant that even its allies and partners felt secure enough to dramatically cut back in their naval investments. In 1986 NATO member states fielded 236 frigates, with 113 of them American. Frigates, being the platform of choice to conduct anti-submarine warfare, were viewed as critical to the alliance, enabling it to counter Soviet submarines in the Atlantic and to escort the convoys of vital men and equipment from the United States to Europe in the event of a Warsaw Pact alliance invasion of the West. During the Cold War, the alliance could also put nearly 200 attack submarines to sea (with the United States supplying over 100 of them) to conduct anti-ship and anti-submarine operations.

Following the Soviet Union's demise, however, Europe chose to focus inwardly on social issues while dramatically cutting defense programs. By 2017 NATO nations, absent the United States, fielded just 51 frigates and 32 submarines.[73] Other key enabling capabilities, such as icebreakers, minesweepers, and military sealift ships, also saw shocking declines. European leaders made repeated statements that the world had fundamentally changed, that threats both on land and at sea had dramatically decreased to the point of irrelevance. For them, notions of security needed to be reconsidered in terms of the social challenges facing the broader European community.[74]

As long as the United States and its treaty allies possessed a combined naval force of overwhelming size, they deterred other powers from competing at sea. However, 2015's US fleet of 271 ships along with Europe's shrinking naval force represented its smallest combined continental force since the English-Dutch wars of the seventeenth century. Many analysts and policy makers will swiftly (and correctly) point out that each ship in the modern US Navy was much more lethal than its counterpart in the World War I era, and certainly a single modern European frigate would make splinters of Admiral Michiel de Ruyter's entire early Dutch fleet, but such arguments cannot counter a fact that the United States' competitors understand all too well: *Even the most lethal ship cannot be in more than one place at a time.* It also ignores the fact that competitors are improving the quality of their ships even as they expand their quantities.

These facts present a problem for the United States and an opportunity for those who would make themselves its enemy. The United States had made commitments to maintain ships forward deployed in key regions as a means of providing security through naval presence, but its shrinking fleet rendered these commitments unmeetable. It also strongly implied that if US ships were to be lost in battle, it would be difficult to replace them, an important component in any would-be enemy's deterrence calculations. These arguments lie at the core of the current maritime debate about the future of the free sea.

The US Navy established its deployment commitments at a time when it had a fleet twice as large. In addition, it set up a maintenance-training-deployment cycle that allowed it to keep its ships in good material condition and its crews highly trained and combat ready when they sailed out of their home ports and set a course for their deployment stations. With a large fleet this resulted in an inter-deployment training cycle, which had one-quarter of the Navy's ships in maintenance, one-quarter in pre-deployment training exercises, one-quarter either going to or coming home from distant deployment stations, and one-quarter actually on deployment. This system worked and it even allowed the regional combatant commanders to gradually increase their requirements. If the European commander needed a few destroyers to do a passing exercise in the Barents Sea on an annual basis or the Pacific commander wanted a carrier strike group to conduct a freedom of navigation operation in the South China Sea, there was no problem. The Atlantic and Pacific Fleet staffs would simply write these exercises into their plans based upon available ships, and so these requests grew over time. But as the fleet drew down, the number of requested ships could not keep pace with the contraction and the ratio of deployed ships to ships in maintenance, training, or transit diverged from accepted norms.

At the height of the Cold War the fleet had around 600 ships and it kept 150 forward deployed. During the post–Cold War era the fleet fell to less than 300 ships. Had the fleet rigorously adhered to its commitment to ship maintenance, crew training, transit, and on-station time, then the number of forward-deployed platforms should have fallen to around 75 ships, but it did not. Due to the priority that regional combatant commanders placed on naval presence, and the desire of Navy commanders to demonstrate the "can do" character" of the service, the number of ships underway only fell to around 100 ships, which meant that at any given moment a third of the battleforce was underway. It also meant that Navy commanders at home were no longer meeting all the presence requests of the regional combatant commanders. In fact, they have been routinely 50 ships short of them. Given that recent naval deployments have not gotten shorter (they actually increased from six months to eight months on average over the past decade), this meant that either ship maintenance or crew training had to be diminished to meet the Navy's presence commitments. Evidence suggests that both were cut.

Ships in maintenance presently in 2020 take longer to repair due to the backlog of problems not addressed during previous shipyard availabilities. These longer yard periods place a greater burden on the rest of the fleet, causing the material readiness of those ships that can deploy to degrade even more quickly. This is otherwise known as a material death spiral. Similarly, shorter periods of crew training were cited in the after-action reports of the three collisions and one grounding that occurred in 2017, the Navy's worst year in recent memory.[75] Seventeen Sailors were killed and two ballistic missile defense capable destroyers were taken out of the fleet for three years for repairs at the cost of hundreds of millions of dollars.[76] Ironically, due to the historic backlog in ship maintenance, the three-year repair period for each of the ships damaged by collisions represented more time than it took to build, launch, and commission the two destroyers originally.

Given that only 100 ships are underway versus the previous Cold War average of 150, and that those 100 ships are in materially worse condition, not fully manned, and the Sailors are not completely trained, then it can be understood that the United States is not as "present" as it has been in the past, and thus is receding on the international stage. This change was easily noted. Regions that had previously experienced two-week gaps between ship visits now went two, and then three, and sometimes six months without seeing a US Navy ship. Seas that had been frequented by aircraft carrier strike groups found themselves making do first with cruisers, then destroyers, and then later the modern corvette, the Littoral Combat Ship. Exercises with allies and partner nations were shortened to limit the amount of time that US Navy ships needed to participate, simplified to allow destroyers or Littoral Combat Ships to replace carriers and cruisers, or outright cancelled in key regions. Other types of crucial US Navy operations

— such as Freedom of Navigation Operations (FONOPS), Transit Passage Operations, or Innocent Passage Operations, which sought to reject excessive territorial claims, illegal baselines, or internal waters claims that endangered the concepts of free navigation or free seas — were conducted less frequently or not at all, in critical areas.

Thus the center of the global international construct of a free sea, upon which free trade flowed, began to crumble and rising powers, who were Eastern in their philosophical culture, continentalist in their viewpoint, and authoritarian in their approach to governance, began to sense the weakness in the West and its champion, the United States. The smaller size of the American fleet, its declining material condition and combat readiness, its increasing failure to maintain forward presence, and the shrinking size of its ally's navies convinced nations that had once been overawed by American naval power that perhaps now it was possible to compete with the world's only "superpower" at sea. A vacuum, created in weakness, offered an invitation to be filled. That this invitation arrived just as other critical political, economic, demographic, and technological factors constructively combined to present rising powers with the opportunity to shift their geostrategic focus from the land toward the sea.

Continentalist Sea Powers and a Closed Sea Strategy

continental power is a nation that is focused on land-based strategies, to include in the modern era land-based air power. One might also add that the two varieties of states — continental and seapower — can be determined by an examination of their history, culture, and art. As the noted British naval historian Andrew Lambert has explained, if a nation defines itself through conflicts and competitions with adjacent land powers, placing more emphasis on its territorial borders rather than its oceanic boundaries, then it is a continental power. Conversely, a nation can possess sea power in terms of large naval and merchant fleets and even use sea power to carry out its political, diplomatic, and economic aims, but still not be a "seapower" if its primary strategic goals are still conceptualized in terms of land territorial control or conquest. If, however, a state possesses a large navy and merchant fleet, its art and cultural memory is dominated by the oceanic trade, and its strategic goals are expressed in terms of controlling sea routes, avoiding entangling alliances, and not getting involved in long land wars, then that state is probably a "seapower."

There have been many continental powers over time. Humans, after all, live on land. There have been but few seapowers; Tyre, Carthage, Venice, and Great Britain are on a very short list, but they are the nations that established and then evolved the global laws and norms that essentially define our modern, liberal, free-trading, capitalistic world. Today, however, the great powers — the United States, China, and Russia — are continental powers, and among them only the United States can claim to have once been a seapower. Additionally, the United States seeks to uphold the established global norms. [77]

Russia and China have been historically focused on (if not consumed by) threats along their land borders. In fact, both have grown through time by expanding their territories both to absorb external threats and to create defensive buffer space for their vital cultural and political centers. Today on land they face each other across a long Asian artificial border, and their uneasy partnership has allowed each to look to the sea with unique but cautious perspectives. Historically, the sea has been a conduit of trade but also a channel for new knowledge and ideas. Russia and China are both traditionally insular, imperial, authoritarian, and even autocratic. They each view the sea and the trade as important to their futures, but each also views the oceans as conduits for potential threats, and an environment that needs to be first tamed and then bent to their purposes. Each has subtly encouraged the other in their maritime adventures because, at present, each stands to gain from the success of the other.

As previously mentioned, the Soviet navy reached numbers commensurate with the United States Navy by the mid-1970s, but neither the Soviet naval strategy nor the capabilities of their ships could attempt to compete with the United States on a global scale. Aside from its nuclear attack submarine force, which the Soviets planned to surge into the Atlantic to interdict US forces crossing the Atlantic to bring aide to Europe, most of the USSR's naval strategy focused on defense of its coasts and its nuclear ballistic missile submarine bastions. In fact, much of its land-based maritime patrol and even its bomber force — most of which were equipped with large air-to-surface anti-ship missiles — were also dedicated to this task. It was the Soviets' strategy to isolate the European continent from North America long enough to consolidate a new position in the western region of the continent in the event of war.

Following the Cold War, the Russian navy cast about for survival strategies to keep itself afloat figuratively, and in some cases literally. It found itself sitting on top of a mountain of nuclear-powered surface ships and submarines that had been purchased when it was able to draw upon the combined manpower and taxable income of thirteen communist nations, but when the "Union" split up, most of the financial burden fell upon Russia, which struggled throughout the 1990s to find the cash to provide basic maintenance and training for its vast naval "overhang." In many cases the Russians maintained these ships with just enough men to monitor and maintain their nuclear plants. With the loss of Ukraine, the Russian navy no longer had a shipyard big enough to build or maintain its larger assets, such as the *Kuznetsov*-class carriers. Effectively, Russia's navy was out of action as the millennial year of 2001 approached.[78]

China's transformation began earlier than Russia's. Following the death of Chinese Communist Party founder Mao Zedong in 1976, his successor, Deng Xiaoping, imposed a pragmatic approach to Chinese foreign policy.[79] Having achieved a modicum of internal political stability, he sought to grow China's economy by cautiously opening it to the world and pursuing "socialism with Chinese characteristics," which was a half-hearted attempt to attract outside capital and technological investment in China. Deng simultaneously sought to increase China's security by building a wall, but not against the traditional threat of Mongols or Manchurian's coming out of the north. Deng realized that the two most recent "barbarian" incursions — the Western imperial powers, which had brought opium and unequal treaties during the nineteenth century, and the Japanese, who had invaded and occupied China in the twentieth century — had come from the east, from the Pacific Ocean. Beginning in the early 1980s the Chinese began to focus on a "Near Sea Active Defense" strategy that would allow China to operate safely and effectively within the first island chain (Ryukyus, Taiwan, Philippines) that lay off the Asian continent. This strategy sought to provide the Chinese with dominance in this vital area, while denying the entry of the United States Navy and its allies long enough for China to be able to land an amphibious force and take control of Taiwan.[80]

Deng's perspective was that of a traditional continentalist. He sought the economic advantages of the open, global system, yet he feared the ideas and change that might accompany allowing it into his nation. He also had the perspective of a leader looking from the land to the sea and seeking a method to control or "close" the sea at a time, and for a duration, of his choosing. Jiang Zemin, Deng's successor, continued this continentalist approach in a 1990 essay in which he asked the People's Liberation Army Navy to "construct the motherland's great wall at sea."[81] The first step in this process was the modernization of the People's Liberation Army Navy, which, as its name implies, really wasn't a Chinese navy at all but rather the naval arm of a land army, and hence forever structurally subordinated to a continentalist viewpoint. Jiang's modernization included the addition of new classes of frigates and patrol craft equipped with modern anti-ship missiles, as well as advanced diesel submarines and land-based aircraft. These platforms were characterized by small armaments and short ranges. Within this strategy, China continued to grant pride of place to the People's Liberation Army in its defense strategy. The central organizing idea was that China would establish an umbrella of shore-based anti-ship missiles and bomber aircraft under which their smaller naval combatants could operate while escorting amphibious assault craft to Taiwan and attacking American naval ships.

The historical waves that emanated from a series of events during the 1990–2001 era combined in a constructive manner to alter Russia and China's approach to the sea and the West. The first "splash" in the pond of global affairs was the overwhelming victory of the United States and its allies and partners in the Middle East in 1990–1991. The turbulence associated with the first use of stealth aircraft in battle along with hundreds of cruise missiles and precision strike weapons convinced both China and Russia that a new age of warfare had dawned. The next stone dropped with the decision to send an American "supercarrier" with its complement of eighty-five aircraft equipped with precision strike munitions along with a light amphibious carrier and assorted escort ships through the Taiwan Strait in 1996 in response to Chinese attempts to intimidate nationalists on the island prior to Taiwan's presidential election.[82]

The last wave maker was the actions of the United States, in cooperation with its NATO allies, and despite the loudly expressed wishes of Russia, against the government led by Serbian Slobodan Milosevic in the late 1990s. This campaign, coinciding as it did with the United States' continued pressure campaign against Iraq's Saddam Hussein, convinced both Russia and China that the United States and its stated strategic goal of expanding the number of nations operating under a liberal capitalistic system of governance was an impediment to their own long-term goals of gaining first regional and then global dominance. The subsequent accidental bombing of the Chinese Embassy in Belgrade with a precision strike weapon dropped from an unseen B-2 stealth bomber in 1999 did nothing to reverse China's and Russia's conclusion that the US would be a long-term obstruction to their strategic aspirations.[83]

Russia and China each moved along similar lines, believing that the key to the United States military's success in the Middle East and the Balkans was the mistaken decision by its opponents to allow it the time and a permissive environment to build up its forces. Saddam Hussein did nothing during the six months it took for the US to move over 500,000 Army and Marine troops into Saudi Arabia. Similarly, Milosevic took no action against allied forces flying from Italian air bases or against US nuclear aircraft carriers operating just off his shores in the Adriatic Sea. Russia and China decided that they would not make the same mistake and each sought to create multiple layers of weapons that could push their enemies back far from their boundaries or shores, creating defensive buffer space while denying US military forces the ability to close within the effective range of their weapons. These layers began with short-ranged fighter and attack aircraft, surface-to-surface and surface-to-air missiles, and smaller surface warships as well as small diesel submarines capable of operating in shallow and confined waters close to shore.

The next layer is composed of long-ranged bombers as well as medium-ranged surface-to-surface cruise missiles and surface-to-air missiles capable of threatening ships and aircraft as they approach. Finally, both nations developed a new generation of weapons to hold US forces at risk as much as 1,000 miles from China's or Russia's shores. This approach is commonly referred to as an anti-access, area denial strategy in the West but could just as easily be referred to as a buffer-layer strategy in that it seeks to hold off a majority of an attacking force's assets and to attrite or destroy those who seek to penetrate the increasingly dense defensive layers.[84]

In addition, by conceptualizing it as a buffer strategy, it more closely aligns with China's and Russia's historical strategic approach, as both nations were built over time through expansion by conquest of territorial rings acquired to place distance between enemies and their strategic central cores. Thus, for China and Russia a modern buffer-layer strategy at sea can be seen as an expression of their long-held continentalist perspectives and an attempt to challenge and alter the principles of the free sea and free trade to align with their governing beliefs and priorities.

Russia, for its part, was aided in its military efforts by a 2001 rise in the prices of petroleum and natural gas energy, which the nation possessed in abundance, and the elevation of Vladimir Putin, the former intelligence officer who had written his doctoral thesis on the need to direct national control over Russia's natural resources, to leadership as prime minister and then president in 1999, and then prime minister again 2008, and then president again in 2012, seemingly for life. Putin possesses a strong nationalistic instinct and a strategic mind that was, and is, capable of making the best use of the limited opportunities afforded him.[85]

Under Putin, the Russian navy stabilized older ships, began to modernize and upgrade a select few platforms that were in the best material condition, decommissioned many that were beyond repair, and finally made significant research and development investments in a select few exquisite technologies that promised great returns.[86] Out of these efforts, Russia began to field a new generation of advanced anti-air, anti-surface, and hypersonic missiles that challenged or exceeded their Western counterparts. Russia installed new weapons in key strategic locations in St. Petersburg, Kaliningrad Crimea, Georgia, and Syria, where their ranges and modern lethality gave Putin an option to erect a modern "iron curtain" that could divide Europe at a moment of his choosing.[87]

Russia had gained access to Crimea and Georgia by deploying "little green men" in "gray zone" strategies that destabilized local governments, allowing Putin's forces to illegally occupy them before outside powers could directly attribute the initial actions to his government.[88] Putin's actions in Syria were more overt and direct but the result was the reestablishment of Russian interests on the eastern end of the Mediterranean Sea.[89] In the end, Russia possessed the capability to push American naval power from the Baltic, Black, and eastern Mediterranean seas at a time of its choosing.

Eastern Europe is not the only maritime environment that attracted Putin's notice. The only effective path to figuratively and literally divide the North Atlantic Treaty Organization was to place control of the Atlantic in doubt, forcing the United States (and Canada) to fight their way across to Europe, should Russia threaten or invade a European NATO member. Even with rising national income, Russia could not afford to recapitalize all of its military forces, to include its navy.[90]

Thus, in order to accomplish its goal, Russia began to invest in a new generation of quiet and fast nuclear attack and cruise-missile-launching submarines that are capable of operating at great depths. During the 2010s, Russian submarines began to operate in limited numbers for short periods of time in the North Atlantic, Mediterranean, and even off the East Coast of the United States. These deployments raised increasing consternation among NATO nations, troubled as they were about the implications of Russia's potential to shut down transatlantic sea traffic.[91] The alliance moved to reestablish an Atlantic facing anti-submarine command, and within the United States Navy, with sources suggesting that the service's anti-submarine skills had fallen off in the decades since the Cold War's end, increased emphasis was placed upon undersea warfare as a core mission.[92]

Perhaps Russia's most important moves with regard to the future of free trade and the free sea are its recent military investments in the Arctic, its legal claims over both the resources of the region, and its assertion of an "internal waters" claim over the newly vital Northern Sea Route shipping lane, which recently opened due to global warming conditions. Russia's wisdom is rendered even more evident by the fact that the US Navy, which would be the entity normally called upon to uphold free sea claims in a maritime region, is materially ill-equipped to operate in the Arctic Ocean. The Navy has not built an "ice hardened" ship since World War II, nor does it possess icebreakers, and its population of nuclear submarines rated as "ice hardened" is about to precipitously decline with the impending decommissioning of the Los Angeles (Improved) class of submarines. Even the US Coast Guard, which is not a military service so much as a law enforcement entity falling under the Department of Homeland Security, at present possesses only one heavy and one medium icebreaker, each of which are unarmed. While the Coast Guard is building a new class of heavy icebreakers and plans to add three to its inventory in the coming years, Russia fields over 40 icebreakers, including several nuclear-powered "heavy" ships as well as a new class of armed arctic rated vessels.[93]

Recently Russia indicated that it plans to invoke Article 234 of the United Nation's Convention on the Law of the Sea, which governs ice-ladened seas, and require all commercial vessels transiting along the Northern Sea Route to be escorted by a Russian icebreaker and to pay a "piloting fee," thus altering the region's character as a "free sea." At the same time Russia is requesting that its Arctic baseline boundaries, which determine its territorial and internal waters claims, be amended to include several new offshore island features. The proposed baselines would have the effect of placing much of the Northern Sea Route within an "internal waters" definition, transitioning the new oceanic route to something akin to the United States' Mississippi River. Russia is also equipping several land bases, both on the Asian mainland and on offshore islands it controls, with advanced anti-air and anti-ship missiles, giving it the ability to control who enters the region.

The net effect of these actions will be to convert a vast portion of the Arctic Ocean into Russian territorial waters. These legal moves in the maritime environment are accepted within the confines of current international law, and are of great interest to Russia's partner, China, which is considering several "adjustments" to the maritime domain itself.

Despite Russia and China both being authoritarian-autocratic governments and the similarities in their current geostrategic approaches to the globe, China's circumstances are entirely different than its northern neighbor's. China's current seaward focus had its origin in the mid-1990s when China was confronted by the aforementioned United States 1996 carrier presence in the Taiwan Strait which also coincided with the nation's shift from being a net exporter to a net importer of energy. Over the past 25 years, as China's economy grew and its bicycle-based transportation system shifted to an automobile society, demand for oil rose exponentially from 2.3 million barrels of oil per day in 1990 to over 13.5 million barrels per day in 2017.[94] Such a shift represented a major challenge to China's traditional economic outlook. Historically the nation had avoided dependence upon foreign imports, always seeking to base its economy on domestic resources over which the nation could exercise exclusive control.[95]

Increasing levels of oil and other raw resource imports, including so-called rare earth elements, as well as the central importance of China's export of manufactured goods to the world market, have combined to make the sea lines of communications essential to the Chinese Communist Party, and thus a growing threat to the its survival. Because its population remains, at least at present, too poor to consume the finished products it manufactures, the nation is also dependent upon wealthy economies in Europe and the Americas to buy its exports. Chinese Communists are also dependent upon coastal export manufacturers to generate the income that it ultimately redistributes inward, and this coastal economy requires a stable maritime environment both to supply it with energy and to transport its goods to foreign markets. They cannot simply trust in the system to provide, especially now that the demand for raw materials to feed and fuel China's industries has so far outstripped domestic supplies. Any disruption would threaten the Chinese economy in general and the Communist Party's leadership in particular.

China's growing economy, especially along its long Pacific coastline, has stimulated the growth of a shipbuilding industry, which now ranks among the largest on the planet, and thus it has not been difficult to divert some portion of the nation's tax revenues and its shipbuilding industrial capacity toward the creation of a modern navy. Over the past two decades, the People's Liberation Army Navy has grown not only in size, but also in terms of its capabilities and missions. The reasons for these investments will be discussed in depth later, but suffice to say that in recent years, the PLAN has begun to gain experience as a "blue water" navy that is capable of operating away from home ports and has begun to venture into maritime regions that in recent decades had been the exclusive operating areas of the United States, its allies, and partners.

The security of China's sea lines of communication undergirds the highest level of the government's interests, its legitimacy, and continued survival. The PRC leadership has pursued over the past generation exclusive and unconstrained rights to sources of raw materials and energy in Australia, the Middle East, Africa, and even into Latin America. It was this new factor, China's dependence on unfettered access to overseas resources, that led President Hu Jintao to direct the Peoples Liberation Army Navy to pursue "new historical missions" after ascending to power in 2004. Two of these new stated missions were defending China's expanding national interests, and safeguarding world peace while promoting mutual development. Both of these mission sets represented a divergence from previous Chinese foreign policy statements in that they implicitly demonstrate that China's leadership recognized that the survival of the Communist Party was tied to the nation's growing economy, which is in turn tied to the global world economy.[96] Beginning in October 2004, Chinese military strategists began to move away from the language of "distant sea defense" and toward a new focus on "far-seas operations."[97]

In December 2008, the PLAN deployed three warships for the first time to the Gulf of Aden to support counter-piracy operations. This was the turning point that nearly everyone missed. Most international officials felt that China's participation in the patrols represented a positive turn toward China taking a role as a responsible actor and helping the international community.[98] However, at least one senior strategist thought it signaled a movement toward outward expansion.[99] This patrol, which has been renewed continuously in the twelve years that have followed for intervals that increased from 120 days to 6 months, is significant for several reasons.

First, it demonstrated the perceived rising threat of piracy to Chinese trade interests. Second, the region encompassed by the eastern horn of Africa (and the African continent beyond it) and the abutting Middle East represent an area of increasing economic interest to China. Third, the subsequent deployments have demonstrated growing maturity in the PLAN's handling of logistic support as well as the critical role port visits play in modern naval operations. Finally, these patrols have led China to explore relationships that have allowed them to establish shore facilities in other nations to provide rest for their

crews, repairs of basic material problems, and replenish-
ment of basic supplies (food, spare parts, and fuel).

This change in focus was evident in a series of strategic port facilities that China began to invest in in the
South China Sea, the Indian Ocean, and the Arabian
Gulf regions. These ports, which were initially known
as the "String of Pearls," were located in Burma, Bangla-
desh, Sri Lanka, Mauritius, and Pakistan and were originally completely commercial in character and creating an
increased diplomatic relationship between China and the
host nation. However, in some cases these host nations fell
behind in their payments to China for local infrastructure
improvements, which led to a change in the relationship.
In other cases China, with the consent of the local government, subtly began to alter the character of the port to support Chinese naval ships.[100]

In many ways, 2008–2009 was a turning point for
China. Its ability to survive the massive recession in the
global economy gave it (and many other intellectuals
worldwide) increased confidence in its authoritarian,
centrally controlled, command economy.[101] China's leaders began to believe that their time had come, and shortly
after the economic crisis the nation's attitude toward the
international system began to change. Gone was the advice
of Deng that China should "bide its time." Instead, discussions began to focus on China's "century of humiliation"
and the need for China to return to its central place within
Asia and the world. In taking these actions China revealed
that it had "weaponized" cooperation with international

actors, using the bumper sticker watchwords of the West to bend the international system to its advantage. The nation's next president, Xi Jinping, who took office in 2013, epitomized this approach in his public statements and in his strategic approach to the world. He expanded the nascent "String of Pearls" investment into the far-reaching Belt and Road Initiative. The "BRI" pursued infrastructure investments in port facilities around the world (the Belt) as well as major rail and highway building projects through central Asia (the Road) to better link China with Middle East and European markets.[102] Perhaps harkening back to China's late nineteenth century collapse under pressures from imperial European powers, China pursued a global trading network that mirrored the pattern laid down centuries before by the Dutch and British East India Companies, but in reverse, and it began to build a new innovative approach to the oceanic environment that suggested a maritime strategy with Chinese characteristics.[103]

Almost immediately following the 1996 Taiwan Straits incident, China approached Russia and purchased five modern *Sovremenny*-class destroyers. Each of these ships came equipped with eight highly advanced and lethal SS-N-22 Sunburn missiles.[104] In addition, the PLAN developed a 60-ship group of *Houbei*-class high-speed patrol craft equipped with indigenously built YJ-83 anti-ship cruise missiles.[105] Through these investments China created a formidable sea denial force. As was their pattern, shortly after purchasing the *Sovremennys*, China reverse-engineered the hull designs and combat systems they found and began to produce their own air-defense-capable *Luyang*-class guided missile destroyers. With phased array radars, these vessels are China's answer to American Aegis air and ballistic missile defense technology. Equipped with the Russian- and Chinese-built surface-to-air missiles, the ships can defend themselves against attacks by modern aircraft. Later integration of longer-ranged anti-ship cruise missiles enabled them to move out to sea, severing their dependence upon shore-based air defense systems to take up distant missions.[106]

China's latest surface ship, the 10,000-ton Type 55 *Nanchang*-class, appears to have 112 vertical launch cells, placing it on par with the US Navy's advanced *Ticonderoga*-class cruisers, which have 122 launch cells. China has built ten of these ships and is poised to build more.[107] China has stated that these ships will not only be equipped with surface-to-air missiles but also the 500nm CJ-10 land attack cruise missile.[108] The *Nanchang*s, with their air defense as well as power projection capabilities, are critical to China's efforts to take the next major step in the evolution of their navy, the development of aircraft carriers.

Aircraft carriers had been in operation for nearly a century when China began to develop its own capability. The British constructed the first full-flight-deck aircraft carrier, the HMS *Argus,* in 1918.[109] Aircraft carrier design and missions evolved over the succeeding century, perhaps reaching their apogee during World War II when the carrier battles at Pearl Harbor, Coral Sea, and Midway were historically decisive.[110] During the Cold War era, carriers, and American carriers in particular, took on other, more symbolic roles, becoming great ships of state, carrying the full prestige of the nation with them wherever they went.[111]

China signaled an interest in carriers in 1998 when it purchased the *Kuznetsov*-class carrier *Varyag* from Ukraine. For years nothing was done with the hull, but in 2005 it was towed into a drydock at the Dalian Shipyard and began a refit, emerging with a new name, the *Liaoning*, to begin conducting flight operations in 2009.[112] While Chinese leaders asserted that the carrier was needed to strengthen China's comprehensive power, it was just as true that China sought to develop a carrier because great powers had carriers, and China was bent on establishing itself as a great power.[113] Carriers came in handy for the Chinese, if even just to overawe their neighbors with a prestige naval asset.

For professional observers, however, the real test of China's naval modernization lay in its development of its submarine force. During the post–Cold War era, China designed, built, and launched the nuclear-powered Type-93 *Shang*-class fast-attack submarines and then the Type-94 *Jin*-class ballistic missile submarines.[114] There is currently speculation about the imminent appearances of the Type-95 *Sui* SSN and Type-96 *Tang* SSBN classes as well.[115]

The Chinese have 12 nuclear submarines. In addition, China also has the air independent diesel propulsion *Yuan*-class and Russian-exported *Kilo*-class diesel attack submarines. Taken together, these ships provide China with a formidable and modern submarine force.[116] These stealthy platforms, equipped with missiles and torpedoes, can operate unobserved for days underwater without surfacing, menacing other naval forces in the region. China currently has about 55 diesel boats in its inventory, with plans to increase that number to 75 in the near term. This diesel submarine force can operate very effectively within the first island chain, freeing the growing nuclear-powered force to press outward, taking on new missions.[117] Today, PLAN far surpasses the United States Navy in terms of size, fielding well over 350 ships in its battleforce, but, true to China's continentalist tradition, the PLAN doesn't actually serve as the front line of China's new "great wall of sand."[118]

China's new great wall moved from a strategic construct to a physical form in 2013, when China began a series of dredging operations that have created over 3,200 acres of new land to build seven artificial "features" in the South China Sea, which include the large constructs at Fiery Cross Reef (677 acres), Mischief Reef (1379 acres), and Subi Reef (976 acres). All three of these dredged-up features now support 9,000-foot-long runways, hangars, and support buildings as well as piers for military logistical support ships to dock at. These features lie within either the territorial waters or the extended economic zones (EEZs) of other nations.[119] It was China's strategic intent to gain recognition of these features as "islands" and Chinese territory with accompanying expansion of their territorial waters and EEZs, but the Philippines, which held original claim over some of the reefs as lying within their own EEZ, filed suit in the Hague International Tribunal, despite Chinese attempts to bully them to initially not file the suit and later to withdraw it.[120]

China's legal assertion before the Court was that nearly the entirety of the South China Sea was China's internal territorial waters, basing this claim upon boundaries defined by eleven dashed lines that appeared on a post–World War II Nationalist Chinese map.[121] Two of these lines that extended into the Gulf of Tonkin were removed out of deference to Vietnam, resulting in the now famous "Nine-Dashed-Line" base claim. An additional dash was later added to extend around Taiwan and into the East China Sea.[122] All these claims did was reveal the lack of historical backing for the original map and the arbitrary nature of China's subsequent extensions, so it was not surprising in 2016 that the Tribunal ruled conclusively against these claims.[123] China immediately dismissed the ruling, demonstrating their willingness to operate outside of international law when it is found to be in opposition to its interests.

Not surprisingly, given its "lead from behind" foreign policy, the Barack Obama administration chose not to offer military or diplomatic support to the Philippines after the Hague ruling was released, despite the fact that the island nation was the United States' oldest treaty ally in the region. This was a repeat of its decision in 2013, when China forcibly occupied the Philippines Scarborough Shoals, not to "take sides" in territorial claims disputes in the South China Sea, and this remained official US policy until July 13, 2020, when the Trump administration decisively rejected all of China's excessive claims in the region, including its claims over any of the dredged locations upon which China built artificial features.[124]

China continues to assert its sovereign ownership of the South China Sea, and further emphatically states that the region represents a "Core Interest" of the Chinese Communist Party, placing it on the same level of concern as China's control of Tibet and Xinjiang, two other regions where China's sovereignty is also questioned. Today the artificial islands and the installations that have been built upon them represent, at a minimum, the potential for a massive expansion of Chinese influence eastward. As one US Navy commander pointed out in 2016, the installation of advanced Chinese aircraft and weapons systems on the features could occur "overnight." These systems could include the HQ-9 surface-to-air missile, the YJ-62 anti-ship cruise missile, and potentially the advanced J-11 attack fighter.[125]

Anti-ship cruise missiles, launched from land, aircraft, and ships, have been recognized as game changers in naval warfare ever since the Israeli destroyer *Eilat* was sunk by an Egyptian cruise missile during the 1967 conflict between those two nations. Additional ship losses during skirmishes between India and Pakistan, as well as the notable loss of HMS *Sheffield*, a then-modern Type 42 British destroyer, and the container ship HMS *Glamorgan* to Argentinian-launched cruise missiles, demonstrated the continued evolution and increased lethality of these weapons over time. China's inventory of these new weapons and their accompanying sensors and command and control nodes potentially installed on its new South China Sea features represent critical components of its new great wall, but not all components of the wall are to be found at sea. In the past two decades, China created two new anti-ship ballistic missiles that have significant strategic implications as well.

In May 2009, the American public first became aware of the DF-21D anti-ship ballistic missile, which quickly became known as the "carrier killer." Of course, the missile had been in existence for some time, going through development and testing, and was only reaching its initial operational capability in 2009 when news of its existence became public. It is a derivative of the DF-21/CSS-5 solid rocket-propelled, medium-range ballistic missile. But what made this 900-nautical-mile weapon different was the inclusion of a maneuvering reentry vehicle (MarV) warhead that allows the kill-vehicle to alter course and maneuver after coming over its ballistic apogee. There is also more than a suggestion that the warhead possesses a seeker that can spot its target and steer precisely toward it despite the hypersonic speeds the vehicle would be traveling.[126] Although the missile's conventional warhead is not thought powerful enough to sink a US carrier, there is concern that it could destroy the carrier's radio and radar antennas as well as damage or destroy any aircraft caught on its flight deck at the moment of impact, thus achieving a "mission kill" of the prestigious platform, knocking it out of action and forcing its return to a repair shipyard in the United States for a significant period of time.

The US Navy was just beginning to conceptualize how to counter the implications of the DF-21D when the People's Liberation Army Rocket Force rolled out the DF-26 "Guam Killer" missile in a September 2015 parade. The DF-26 is thought to have a maneuvering warhead similar to the DF-21D, but sitting upon a larger rocket booster, allowing it to reach the island of Guam and potentially taking out aircraft parked in hangars or ships tied up at piers there.[127] These two new missiles represent a serious attempt by China to force the United States to question the survivability and effectiveness of its multi-billion-dollar aircraft carriers, thus isolating the Chinese mainland from their disruptive, regime-changing attacks.

It is necessary to understand the full implications of Russia's and China's recent actions and investments from a geostrategic perspective. Both view the seas and all the activities that occur upon them, from commercial to military, as threats to their national interests. Both nations need to either import or export raw and finished goods upon the sea and hence have become critically dependent upon it as a medium for economic growth and even survival. Both desire access to fishing and mineral resources that lie beneath the surface of the seas near and around them. As authoritarian states, both fear the sea and the ideas it transmits — namely free trade, free navigation, and, by extension, individual freedom and liberty — as threats to their internal stability. As such, over the past three decades they have sought to formulate, impose, and gain international acceptance of a new strategic approach to the sea that is, in many ways, diametrically opposed to the maritime principles that have sprung from the intellectual tradition of the Western enlightenment. Whereas Western theorists Karl von Clausewitz, Halford Mackinder, and Julien Corbett viewed land as the decisive theater of war with regard to competition between nations, China and Russia, as land powers, have sought to "offshore" the potential for conflict in order to guarantee a local *Pax Terra*, thus maximizing their internal stability. Also, increasingly, the economic vibrancy of the sea has made it the "heartland" of global trade and hence the location of any future "pivot" vis-à-vis great power competition.

This would tend to suggest that China and Russia have fully bought into Alfred T. Mahan's oceanic view of the world, but such an understanding would be incorrect because both China and Russia do not believe that victory can be achieved at sea via fleet versus fleet decisive battle. Rather their entire "naval" strategy is predicated upon the premise that land-based sensors and weapons, including fighters, bombers, cruise missiles, and ballistic missiles, can provide a defensive umbrella that their maritime assets — naval ships as well as commercial merchant vessels — can operate with impunity. In addition, they believe that, in wartime, these same land-based systems can combine with sea assets to defeat an enemy fleet far from their shores. Again, some have used the term "anti-access and area denial" to describe China's and Russia's investments in the western Pacific and the Arctic oceans, investments that seek to create buffer-layers to keep US forces focused on regime change far from their coasts. This may well have been the original intent, but recently both nations have begun to articulate a transformative change in maritime governance that seeks to create and gain acceptance of a *mare clausum,* a closed sea, while also advancing incremental expansions of their territorial sea claims.

The logic of this argument has always existed within the accepted legal dialogue that created the principle of *mare liberum* to begin with. Like Damocles' sword, the threat of a closed sea always hung suspended by the finest of threads over the concept of a free sea. As mentioned previously, when the Dutch legal scholar Hugo Grotius formulated his argument for a free sea, it was not universally accepted, even in his own country, where another Dutch jurist, Cornelis van Bijnkershoek, argued that a nation's sovereignty did not end at the water's edge, but that territoriality should extend to the limit that control of the sea could be exerted by land-based weapons. As he succinctly stated, "I hold that the territorial dominion ends where the power of weapons terminates" and his words led to the "cannon shot rule," which established a territorial sea "carve out" exception to the concept of a free sea, extending sovereignty initially to three miles and now to the commonly held twelve-mile norm.

Based upon the ability of their sensors to see and their weapons to reach, China and Russia are now seeking slow acceptance of expanded territorial sea claims. Through their "lawfare" arguments and their actions, they seek to acculturate the West to the idea of external seas becoming internal waters. Because the power of their weapons now extends far out to sea, logically their dominion should follow. What is more, through China's "Belt and Road" initiative, it now could extend this concept of sovereign sea control around the world as adjacencies to its international port investments. To be sure, they cloak their claims under the aegis of the United Nations' Convention on the Law of the Sea and find support in contemporaneous "historical" documents and maps, but the result is all the same. They are pursuing a continentalist strategy that utilizes land-based weapons supported by naval ships to close the seas. They do this to guarantee the safe transit of their supplies and merchandise while shielding themselves from kinetic and philosophical attack from an outside liberal, capitalistic sea power.

One should consider the following scenarios. What if Russia were to use its heavy icebreakers or China to use its new Type 55 heavy cruisers, each of which far out-mass the displacements of American Ticonderoga or Burke-class ships, to physically block or "shoulder" American ships out of waters that Russia and China feel are territorial seas? This is a tactic that was used repeatedly by the Soviet Union during the Cold War. What would the United States do? What if these activities were accompanied by the simultaneous activation of fire control radars associated with Russian and/or Chinese shore-based anti-ship missiles? And what if another incident arose like the one that occurred in 2001 when a Chinese fighter aircraft slammed into a Navy EP-3 surveillance aircraft, resulting in the loss of the Chinese plane and pilot, and the emergency landing of the EP-3 on China's Hainan island? How would American leaders react if a rammed American ship took on water and sank? Both Russia and China enjoy superiority of ship numbers in their home waters; how would the United States respond to such aggression? How would we maintain our "face" with our Asian allies and partners?

To avoid these potentialities and defeat Russia's and China's attempts to extend territoriality over their near seas, the US Navy, along with its allies, must rapidly rebuild their fleets and then return to the global oceans in a persistent and credible basis with ships that can be counted upon both to win wars and to maintain the peace. We must reestablish the concept that, yes, Russia and China may well consider initiating a war, but they should not begin one today.

Upholding the Law Means Getting the Right Mix, Playing Ball, and then Finding the Beat

In 2007 the United States Navy published "A Cooperative Strategy for the 21st Century" and one of its first observations was "preventing wars is as important as winning wars," which represents perhaps the most succinct statement of the critical importance of modern naval forces.[128] In peacetime, land forces are in garrison preparing for war. Air forces can surge outward to demonstrate capabilities, but they cannot persist in the critical global common areas where competitions between nations are played out. Only naval forces, operating as self-sustaining platforms that can move from place to place and yet persist within a region, resupplied by long and healthy logistic support line of communication, have the ability both to preserve the peace while also preparing to win the war.

Therefore, both the size and the internal composition of a nation's navy are critical to its ability to operate as a great power. The dynamic equilibrium between winning a war and preserving the peace, and by extension ensuring the survival of the global liberal capitalistic order, must be the focal point of American fleet design.

The primary purposes of a national military are to ensure the survival of the state and to promote its interests, by force if necessary. By extension, a naval force's primary requirement is to win a nation's war at sea by protecting its lines of communication and controlling those portions of the seas necessary to achieve military victory. In other words, there will always be some aspect of a navy that is focused on generating war-winning capacity. Within the United States Navy much of this war-winning potential is lodged in the eleven "supercarriers" that the Navy is required by law to maintain within its force and the power projection and sea control capabilities of their embarked air wings.[129] These carriers represent the central element of that nearly unique US naval force, the carrier strike group.

The other sub-components of these groups are the carrier's escorts, the cruisers, destroyers, and submarines that accompany the "big decks" as they make their deployments, providing defense against enemy submarines, surface forces, and aircraft, as well as against land-based cruise and ballistic missiles. During the Cold War, strike groups were composed of no less than two cruisers, four destroyers, four frigates, and two fast attack submarines. Today, with the shrinking size of the fleet, carriers often find themselves accompanied by as few as three escorts and often are joined by allied ships to make up their defensive shortfalls.[130] While these escort ships do have some offensive potential themselves in the form of land and sea strike missiles, much of their tactical focus is on protecting the supercarrier, the predominant capital ship of the American fleet.

The aircraft carrier has been the centerpiece of the American Navy since December 7, 1941, when the Japanese sank or seriously damaged most of the US battleships in the Pacific during their attack upon Pearl Harbor, leaving the three carriers that were at sea that day to take the load in the months that followed. Later, other carriers were built and dominated the war, even after the battleships attacked at Pearl Harbor were refloated, repaired, and made available to the fleet again.[131]

The ability of the carrier-based aircraft to project power ashore and to achieve control of the seas around the carriers made them the preeminent naval weapon of the war, achieving a draw at the battle of Coral Sea and a decisive victory at the battle of Midway. Following the war, the Navy doubled down on the carrier as the center of its fleet architecture by designing and building the massive Forrestal-class "supercarrier."[132] The *Forrestal*-class carriers, and the subsequent supercarrier classes that followed, were all designed to launch and recover aircraft large enough to carry nuclear weapons and strike targets deep inside the Soviet Union. At present it has ten *Nimitz*-class carriers with one new *Ford*-class carrier fitting out.

These nuclear-powered vessels can operate with 80–90 air-craft onboard, although today's carrier air wings generally include only 65 planes.

Following the end of the Cold War, most of the carrier air wing's long-range aircraft, the A-3 Skywarrior, A-6 Intruder, and F-14 Tomcat, were decommissioned and replaced by short-ranged, light attack FA-18 Hornet and Super Hornet aircraft. Due to the Hornet's limited range, the Navy transitioned to an era where it increasingly operated its carriers close to the shore of its targets, such as in the Arabian Gulf or Adriatic seas, while taking care to surround the carrier with ships focused on defensive "anti" mission sets: anti-air, anti-surface, anti-submarine, and most recently anti-ballistic missile defenses. Arguments have been advanced once again to increase the lethal range of the air wing through the introduction of unmanned strike platforms, but there seems to be little appetite within current naval aviation leadership to incorporate these aircraft within the carrier air wing.[133]

Other proposals have been made to utilize the carriers and their air wings to create protected enclaves in the middle of the ocean where large "big-wing" Air Force tankers could operate, refueling large Air Force bombers such as the venerable B-52 Stratofortress, which can fire standoff hypersonic boost-glide missiles as well as traditional subsonic cruise missiles against continental targets. Carrier air wing aircraft could protect Air Force tankers refueling B-2 and B-21 stealth bombers, enabling them to penetrate deeper into enemy territory and to remain longer, hitting key targets. These aircraft, along with the B-1B bombers, could also conduct war at sea, with strikes against targets the Navy is unable to reach.[134] Such dependence on Air Force assets is in line with World War II concepts of operations, but it also represents the Navy's failure to retain its own long-range, carrier-based strike capability.

Cruisers would be the lead ship in defending the carrier and these protected enclaves. As a class they historically sailed, or "cruised" (thus the name), independently to show the flag in peacetime and raid enemy commerce in wartime.[135] Following the conversions of navies from sail to steel and steam, cruiser designs attempted to integrate their historic roles, which included large guns to attack commercial shipping, but also large engines and little armor, allowing them to outrun larger, heavier, and slower battleships.[136] However, during World War II, cruisers found a new role escorting aircraft carriers, using their array of small- and large-caliber guns to provide defensive fire against attacking enemy aircraft. Following the conflict, cruisers were among the first ships in the Navy to integrate surface-to-air missiles, which allowed them to target attacking aircraft farther from the force and with greater lethality. Today's *Ticonderoga*-class cruisers, built from the mid-1970s through the early 1990s, are equipped with the Aegis Mk VII combat system and 122 vertical launch cells, each filled with either surface-to-air, ballistic missile defense, anti-submarine warfare, or Tomahawk land-attack cruise missiles. The class represents the most lethal, defensive ship in the world today. To fully defend the nation's eleven 100,000-ton supercarriers, the Navy requires a total of twenty-two cruisers to carry out normal operations. The Navy has these today but plans to retire all *Ticonderoga*-class ships over the next decade as they reach the end of their service, well past their originally planned thirty-year lives.

From a historical perspective, destroyers are a relatively new part of the fleet. Whereas "battleships" descended from "line of battle" sailing ships and cruisers draw their name from traditional commerce raiders from the age of sail, destroyers as a concept emerged after the invention of the long-range torpedo, a propeller-driven subsurface weapon that was associated with shore-based, small, fast "torpedo" boats that would drive directly at enemy battle-ships, launch their weapons, and then turn and quickly retreat beyond the range of enemy guns. Such was the concern of battleship admirals about these new pesky enemies that they established a requirement for "torpedo boat destroyers" that had sufficient range to accompany battleships to foreign waters and yet were small, quick, and maneuverable enough to defend the larger capital vessels against enemy attacks. These ships evolved over the past century to become the modern destroyer.

In the post–World War II era these ships were designed with a focus on anti-surface and anti-submarine warfare in and around the carrier. In the modern era, with the appearance of the Aegis combat system–equipped *Arleigh Burke*-class destroyers, this class of ships has also taken on anti-air and ballistic missile defense roles, becoming the modern jack-of-all-trades, with some question as to their mastery of any of them. Each of these ships is equipped with ninety-six vertical launch system tubes, which can carry the same variety of missiles as the *Ticonderoga*-class cruisers.

During the Cold War, each carrier would be accompanied by four destroyers to properly screen and defend it, and this would drive a requirement for forty-four of these ships, but destroyers now also serve in ballistic missile defense roles both for other ships as well as bases ashore. In the near future the Navy will have to decide whether to perform a service life extension on the first flight of the *Burke* class, some twenty-one ships, and some thought should be given to modifying them to carry the new Conventional Prompt Strike missiles now entering service.[137] Such an addition to these older ships would turn them into lethal "three-point" shooters, able to hit critical targets from well outside the range of enemy missiles. These requirements plus presence roles drive a requirement for seventy-three modern guided-missile destroyers.

As the Cold War evolved, both sides made significant investments in submarines, to include independent development of attack submarines, which initially focused on targeting enemy surface forces but later developed the capabilities necessary to locate and target enemy submarines. US submarines are equipped with nuclear reactors, giving them the ability to operate continuously beneath the seas without surfacing to recharge batteries or refresh their air supply. Because of the rising threat of enemy attack submarines against the carrier, it became common for each carrier to be escorted by two of these submarines, but to be clear, most attack submarine missions were independent in character. For instance, these boats were continuously dispatched to north Atlantic and Pacific regions where Soviet nuclear ballistic missile submarines were thought to lie in wait for orders to launch their weapons of mass destruction at the United States. It was the US Navy's job to keep continuous "trail" on these boats and to be prepared to destroy them if it was perceived that they were readying an attack. Other attack boats would be dispatched to lie just off a competitor's coastline, monitoring activities and communications while gathering intelligence of other forms.

During the Cold War, when the United States Navy had nearly six hundred ships, some one hundred of those platforms were submarines of one class or another. They were equipped with torpedoes, mines, and, near the end of the Cold War competition, Tomahawk land attack cruise missiles. Today's Navy is constructing *Virginia*-class attack submarines, which can carry all of these weapons, plus the new hypersonic conventional prompt strike weapons that are entering service during the 2020s. A modern US Navy requires no less than sixty-two fast attack submarines today to provide "war-winning" capabilities to the fleet, but in today's world winning the war is, as they say, only half the battle.

The US Navy also has a significant role in "showing the flag," demonstrating and upholding American national interests, and preserving the global peace. This role can possibly be better understood by viewing it through the lens of law enforcement. Think of it as a "cop on the beat" patrolling a local neighborhood.[138] In the early 1980s political scientists James Q. Wilson and George L. Kelling advanced the "broken windows theory," which postulated that degraded outward appearances, such as broken windows, served as a social invitation to break other windows in the same building or neighborhood. However, Wilson and Kelling offered the observation that increased police presence and attention to small crimes instilled a greater sense of order and in turn invited more adherence to rules and norms, raising the standard of living within neighborhoods. The key to the theory was that increased police presence in the communities, patrolling on foot and not in cars, conveyed the sense of the police being part of the neighborhood and not outsiders just cruising through.[139]

After the terrorist attacks on September 11, 2001, the US military saw its operational pace pick up from a trot to a sprint, and that sprint lasted for the better part of two decades. During that period, one in which the fleet shrank from 316 ships on the day the planes hit the Twin Towers to 271 ships in 2015 as troops began to dramatically draw down in Iraq, the Navy's operational deployments did not. Ships were constantly moving from their home bases on the East and West Coasts to the Middle East and then home again, leaving key maritime regions of the world unpatrolled, and windows began to break. Fishing rights of local nations were trampled as foreign fishing fleets swooped in to net up local stocks. Critical energy infrastructure of allies and partners on the bottom of the sea came under threat from rising great powers, who increasingly advanced the idea that the wishes of larger and more powerful nations should hold sway, regardless of the law.[140] Incidents of piracy increased, first off the horn of Africa, and then slowly they spread eastward into the archipelagic waters of the western Pacific. China's move to illegally claim key features in the South China Sea and then construct its illegal artificial islands by dredging up sand and, more shockingly, grinding up coral reefs without any significant pushback from the global community was another blight upon the neighborhood.[141]

Window after window shattered without any policing power stepping in to enforce international rules and norms, causing the local maritime regional communities to begin to lose faith in the concepts of free trade, free navigation, and the rule of law. To be clear, US Navy ships did pass through these regions on their way to someplace else that was more strategically important, but their impact was similar to a police cruiser passing through and pausing only long enough to roll down the window and look quickly around from behind mirrored sunglasses. The US Navy lost touch with "neighborhood" areas with which it had long-established, national interests.

One of the reasons this happened was that the composition of the Navy had changed significantly during the early 2000s. During the days of a 600-ship fleet, the Navy had maintained nearly 200 ships; frigates, patrol vessels, and shallow-draft, flat-bottomed amphibious assault ships, that routinely made port calls in the myriad of out-of-the-way, small, shallow ports that dominate the world's maritime belt where the sea meets the land. Following the Cold War, many of these ships were decommissioned, replaced by far fewer, but more capable, deep draft vessels who could not comfortably visit the smaller harbors even when there were ships to spare for such visits. The eastern Mediterranean, where the Navy previously routinely dispatched a three to five-ship surface action groups of made up of cruisers, destroyers, frigates, and an unseen fast-attack submarine, today has to make do with ships transiting through its waters en route to assigned stations in the Black Sea or the Arabian Gulf. The US Southern Command, which once hosted large squadrons of surface ships, amphibious ships, and even a large carrier strike group as well as running an annual large regional exercise that involved all partner nations, finds itself today making do with ad hoc and inconsistent ship deployments, most of which are focused on counter-drug operations rather than building critical, unified, military capabilities. The presence of the Navy was no longer felt, and a vacuum of influence began to grow.

Sensing this, China began to expand its navy, largely comprised of smaller guided missile corvettes and frigates, and pressed outward across the first and second island chains, as well as into the Indian Ocean and along the coast of Africa. Russia began to deploy its ships again into the eastern Mediterranean and the Baltic Sea, as well as ramping up operations in the Arctic ocean. Both began to pull into ports that the US Navy no longer frequented, subtly encouraging local actors to pivot away from the United States. Iran, always a troublemaker, increased the belligerence and operations of its small boat crews in the Arabian Gulf and around to the Gulf of Oman and the Red Sea. It even had the audacity to capture two US Navy small boat crews operating in international waters.[142] More stones were thrown, windows broken, rules began to erode, and lawlessness in reference to previously established norms began to emerge.

The ongoing drawdown of American commitments to the Middle East and Afghanistan has presented an opportunity to engage the world once again upon the oceans as a seapower with something more than just overawing, high-end, war-winning forces. The US Navy needs to symbolically get out of its high-end cruisers and begin walking the beat, getting to know the world's maritime regions with low-end frigates, patrol craft, and smaller amphibious ships once again. It is time once again to invest in the "influence" portion of Alfred T. Mahan's *The Influence of Sea Power Upon History, 1660–1783*.

Frigates of various sizes and designs make up a vast majority of the naval ships used by the United States' allies and partners. Serving traditionally as the "eyes" of the fleet by providing presence abroad while operating close to shorelines, modern frigates increasingly took on roles as primarily anti-submarine warfare assets while backing up other ships engaged in anti-surface warfare. Frigates have also served as escorts for military convoys, moving men and material across the Atlantic during the halcyon days of NATO. During the days of the 600-ship, Cold War Navy, the United States possessed over a hundred frigates. Today it has none. While it is at present building nearly thirty new littoral combat ships, these single-mission platforms (they can execute either anti-surface, anti-submarine, or anti-mine missions, but only one mission at a time) have not matured well, and the Navy recently decided to decommission the first four ships in the series after less than a decade of service.[143]

These ships clearly are not frigates. However, the Navy has decided to build a new multi-mission guided missile frigate based upon a proven design derived from the European FREMM (FRigate European Multi-Mission) frigates that are already operating in four navies.[144] As such, the United States has made a decision to position itself as a true peer and partner with many of these nations, who look forward to looking across from their control bridges to a US partner, who will look them in the eye rather than looking down at them literally and figuratively from their loftier perches on Aegis-equipped cruisers and destroyers. These new frigates, being smaller in length, beam, and draft depth, will also be able to return to the more confined ports that the US Navy used to frequent but has largely abandoned over the past two decades. Such visits will rebuild the sense of partnership and trust that has begun to atrophy. Using frigates for these day-to-day "low-end" roles will also have the effect of freeing up larger, more complex ships to focus on maintenance, crew training, and most importantly their high-end, war-winning missions. Given that each of these new ships will come with 32 vertical launch cells, they will also have a role to play in a high-end fight as well. Finally, the new frigates will be purchased at half the cost of their *Burke*-class destroyer brethren, providing the Navy with an opportunity to rapidly grow the fleet toward its stated goal of 355 ships. The Navy initially set a building goal of 20 new frigates but the real number, based upon requirements, should be 60.

Both cost and numbers matter. In modern warfare, the technological systems carried inside a ship evolve more rapidly than the hull design that surrounds them, and most naval ships are expected to serve from 25 to 30 (50 years for most carriers) years depending upon the class of ship. Hence, greater emphasis should be placed on new sensors and weapons, and ships should be designed with both sufficient space and power margins for future growth. With this mindset, the ship conceptually becomes a "pickup truck" that carries weapons rather than being conceived as the weapon itself. Such a change in how we approach ship design allows for a break from the trend toward increasingly exquisite and expensive ship designs that are also fragile in the operational environment and moves toward a more rugged, ubiquitous platform that can evolve over time. Such an approach, which implicitly acknowledges both fiscal reality and the high threat environment, allows for the purchase of more, cheaper platforms.[145]

Frigates are an example of these types of ships, but not the only one. Offshore patrol vessels, which have less range than larger ships but can operate effectively from forward bases in smaller, shallower sea basins such as the Azov, Baltic, and Black and Mediterranean Seas as well as the Arabian Gulf, can also provide much-needed presence and "influence" at a cheaper price, as could some of the smaller amphibious ships now being explored by the United States Marine Corps.[146] The Navy Marine Corps team will ultimately require forty of these new ships.

This approach to shipbuilding and naval force design is not so far removed from the *Moneyball* phenomenon that overtook America's favorite pastime, baseball.[147] Author Michael Lewis in his 2003 book of the same name revealed how modern analysis suggests that it isn't the game's big hitters that rule the ballpark so much as the players who consistently get on base via singles, walks, or being hit by a pitch. Research demonstrated that the importance of being on base, as measured by "on base percentage" is more determinative of team wins than individual batting averages or slugging percentages, and so, during the era of home run hitters like Barry Bonds, small-market teams with lower salary caps quietly invested in the cheaper players who nonetheless demonstrated the ability to get on base more often, because, by being on base, they created the opportunity for the next player, who possessed a similar overlooked on-base ability, to move them around the basepaths and ultimately home for the score and eventual win.[148] *Moneyball* revealed the importance of being present in order to create wins, and it suggested a parallel within the naval environment.

Forward-deployed ships demonstrate consistent interest, but they are not so threatening as to force a confrontation. After all, naval ships are always at sea, traversing a free global common, transiting from here to there, so their appearance near a particular country can be officially construed as neither provocative nor alarming. Naval ships providing presence have an elasticity to them. They come, bearing a nation's sovereignty and power, but then they go away as well. As such they act as a lubricant between the grinding tectonic plates of great powers, allowing them to slide past each other by creating a series of small but persistent shocks. But if those ships disappear for too long, if interests are not demonstrated through frequent, smaller interactions and tensions are not bled off, then those tensions rise and the resultant subsequent movement of great power upon great power interests against each other becomes necessarily larger, sometimes leading to the outbreak of wars. *Pursuit of a larger navy to remain forward deployed acknowledges two fundamental truths that have emerged in recent years: You cannot surge trust, and there is no such thing as virtual presence.*

A nation must remain consistently forward deployed with a credible force if it is to convince a competitor that today is not the day to go to war or even remotely risk that event. Strategically located forward bases, outside of buffer-layer threat rings but sufficiently close to provide logistical resupply, could also assure regional partners.

Forward-deployed naval forces would also provide a sizable portion of the force to both absorb the first blow from an enemy and then to respond with a rolling start into planned war-winning maritime campaigns. Like less expensive players who can get on base more often, cheaper, smaller, and less complex ships like frigates, offshore patrol vessels (traditionally called "corvettes"), and smaller amphibious ships can get on base in contested maritime environments and be more present in the world, thus demonstrating their nation's interests, and executing global competitions in smaller, non-kinetic battles on a daily basis. Thus, if the US Navy maintains its high-end carrier strike groups and nuclear-attack, missile-launching submarines as its "war-winning" core while also investing in low-end frigates, patrol vessels, and small amphibious ships as its "preserve the peace, cop on the beat" force, it will find an equilibrium that can counter the growing number of broken windows while also developing and building a Navy that can win decisively if the global system collapses.

A *Moneyball* approach to naval force structure will be cheaper to build than simply expanding the current "high-end" war-winning carrier strike group/nuclear-attack submarine naval architecture that the nation has today, but it will not cost less than the nation is currently spending. For fiscal year 2020 the Department of the Navy submitted a budget request to the Congress for $205.6 billion, which represented a 4.8% increase over and above the enacted fiscal year 2019 budget. Reaching 355 ships for the US Navy will take still more than this, and in fact, given that the actual number of ships required to fully service regional combatant commander needs falls closer to the 426 ship fleet described by Representative Jim Banks's 2019 "Five Ocean Navy Strategy" or the 450-ship fleet remarked upon by former Chief of Naval Operations Admiral Jonathan Greenert during his March 12, 2014, testimony before the House Armed Services Committee, the Navy will need a lot more.[149] The cost to reach 355 ships and then man, equip, train, and maintain that expanded fleet will be, on average, more than $20 billion per year for the next ten years, resulting in a Navy budget of $225–$240 billion. While this number may seem large, it is right in line with inflation-adjusted Cold War naval expenditures.[150]

Additionally, by Cold War metrics, if the nation were to return to its historic levels of spending on defense as a percentage of its gross domestic product during a time of great power competition, then its 2019 defense budget of $732 billion (3.4% of the nation's GDP) would rise to a defense budget of $1.39 trillion, thus allowing for a 450- or even a 600-ship Navy.[151] This number seems large to even the most cynical observer and undoable when the United States' overall national debt of some $27 trillion is considered. Thus, if the nation is to maintain peace in the world and build a war-winning military force, then changes must be made within the overall defense budget, and such changes should be driven by a reconsideration of the nation's overall strategic approach to the world.

By Land or By Sea?

The nation has reached a moment where it no longer has the resources or the breadth of strategic focus to continue its ahistorical attempt to be both a continental power and a seapower simultaneously. The nation is exhausting itself in its present attempt, and thus rendering itself vulnerable to exploitation by enemies who, although not as technologically advanced, have a clearer vision of the world as it presently is. The state of confusion on the part of the United States is surprising given the number of national interests that lie at sea. It seeks the uninterrupted transportation of goods from suppliers to consumers and the right to mine or extract resources from the sea that can be refined and developed on shore. It seeks the uninhibited movement of information across the seafloor as much as it desires the flow of data up to and through satellites in space. The nation desires the continuation of a free sea and thus its strategic focus should adjust its orientation accordingly.

As previously stated, George Washington remarked in his farewell address as president, "Why forego the advantages of so peculiar a situation?" and his observation is just as true today. Our nation largely finds itself safely upon the North American continent with oceans creating deep defensible boundaries to our east and west, with the additional benefit of sharing land borders to our north and south with friendly powers with which we engage commercially and diplomatically in the most peaceful and productive terms. Washington went on to say, "Why quit our own to stand on foreign ground?" It is a question that rings loudly today following some seventy-five years of overseas obligations, toil, and bloodshed by our nation's land forces. The nation can say that it was willing to shed its most precious blood to promote the spread of its democratic republican principles, but these efforts have only succeeded to a limited degree in some areas and not at all in others.

Now the nation faces two rising great powers that each possess nuclear weapons, a fact that all but negates the option to attack, occupy, and garrison them with land forces in a future war. At most the United States will require its military to project power into the Chinese or Russian homelands in an effort to coerce changes in their policies, or, at least, to quietly stand off their shores and uphold the laws and norms of a free sea, free trade, and individual liberty, which are so inextricably bound up with each other. Finally, the United States must change its orientation back toward the sea, making a conscious decision once again to become a seapower, if for no other reason than that China and Russia, the recognized rising great powers, are already making the disruption of the oceanic global commons the center of their own strategic plans.

China has made a series of strategic decisions to cast its future upon the seas, or at least upon the control of the seas. They have grown a manufacturing sector that is the fastest-growing economy in the world. They have built ships to carry raw materials from the world's mines and wells to their factories and others to carry their products to markets around the globe. They have built a nearly 400-ship navy capable of protecting both their home waters and, with increasing confidence, their important and growing network of sea lines of communications. They understand the intersectionality of these three aspects of their national life: An export economy carried on a domestically built merchant fleet that is protected by a growing navy. Finally, they have invested in new land-based missiles to enforce a new definition of territorial seas.

Russia, with declining resources and national power, has played its poor hand of cards extremely well, investing in a series of exquisite nuclear submarines with the ability to interdict American forces attempting to cross the Atlantic, and in a family of land-based missiles based in key geostrategic locations along the border between Europe and Asia. Putin's government has also built a new generation of icebreakers and armed shore bases and offshore islands to render a good portion of the Arctic "internal waters." Each of these investments have been made with the intention of sowing doubt and discord into the North Atlantic

Treaty Organization, Russia's primary strategic threat, and they have largely succeeded. Each considered their environment, created an appropriate strategy, and executed it with varying degrees of success.

The United States, for its part, has not. As an Israeli politician once noted, "You can depend on Americans to do the right thing when they have exhausted every other possibility."[152] The US has been cavalier about its future since the end of the Cold War, coming nowhere near the deliberative process by which the nation created the containment strategy to deal with the Soviet Union during the 1950s. The Trump administration's 2017 National Security Strategy took a traditional continentalist approach to the world by outlining a plan to protect the homeland, promote economic prosperity at home, generate peace through strength by modernizing and expanding the military, as well as renewing the nation's diplomacy and statecraft. The document suffers from its lack of direct approach to the common theme of threats originating from the three global commons: cyberspace, space, and the oceans.[153] This oversight is repeated in the Department of Defense's National Defense Strategy of 2018, which attempted to divide the department's approach to the world into contact, blunt, surge, and homeland layers, terms that betray a heavy continentalist viewpoint but do not translate into naval (or cyber or space) strategies or tactics.[154]

The United States' numerous experiences with land wars since 1945 have created a set of blinders that prohibit our nation's leaders from correctly viewing the geostrategic setting of the current competition and the necessity of turning back to the sea while placing emphasis on those aspects of its military, namely naval, long-range air and space forces that are critical to maintaining free global commons. These forces — naval, air, and space — must also be allowed to fully integrate their efforts to maintain an open maritime common.

This effort, the preservation of the free sea commons and all the principles that descend from that concept, is the most important task of our time. Our competitors have recognized that it is the source of our strength even as we have forgotten it. Their affinity for authoritarianism and continental hegemony is threatened by the ideas that flow into their countries along with free trade, and they now seek to assert control over those ideas before they so "destabilize" their populations that they become ungovernable, at least from an authoritarian perspective. They also seek assured access to their overseas suppliers, overseas markets, and the lines of communication that connect them without fear of interference or interdiction by the West.

Finally, they seek to push the West far from their shores by assigning sovereign characteristics to their near seas and threatening foreign naval vessels with shore-based missiles in a modern application of the "cannonball rule." Their efforts, if successful, would overturn centuries of maritime precedent and undermine the philosophical foundation of the Western liberal order. The United States needs to rededicate itself to the sea and to becoming a seapower, and that implies so much more than simply rebuilding the Navy. To accomplish this goal, the nation's leadership, both in the executive and the legislative branches, must reach a consensus on a strategy that philosophically, structurally, and economically evolves the way the United States approaches the world.

It needs a new national security strategy that is maritime in its focus. It would be a strategy that shifts its primary focus away from a continentalist focus on the Middle East and Europe and toward the ability to control key maritime regions, choke points, and the sea lines of communications that connect them. It would be a strategy that seeks to sustain the United States as a great power by prohibiting China's and Russia's rise through offshore influence and balancing. It would be a strategy that promotes export-focused manufacturing as well as its data, financial, and service sectors. It would be a strategy that assures American access to, and control of, movement across the global commons such as cyber, space, and the world's oceans rather than seeking the means to capture and garrison territory outside of the United States. It would be a strategy that seeks to preserve the global peace by providing offshore naval, air, and logistical support to allies and partners while they, with their own land forces, seek to protect their interests domestically and along their borders. Finally, it would be a strategy that achieves its ends by asserting persistent and credible presence in those areas of the world where the United States has critical interests without having the nation unnecessarily entangled with continental intrigues ashore. Such a strategy should be supported by accompanying National Defense and National Military strategies that support and align to these key pillars.

From an economic perspective, rededicating the nation to becoming a seapower again also means strengthening its shipbuilding industries and associated downstream parts and material suppliers, to include expanding the nation's steel production industry. In 2020 US Navy shipbuilders are often dependent upon single suppliers of critical components in the United States and, in some cases, some components can only be found overseas. There is also a limited market in the United States for certain types of hardened or armored steel required in the building of warships. These critical vulnerabilities should be addressed as part of a defense industrial base revitalization commensurate with that described in Arthur Herman's book *Freedom's Forge,* which detailed the planning, legislation, and men who produced thousands of merchant and naval ships during World War II, because becoming a seapower again would entail more than just rebuilding the Navy; it would also require rebuilding the nation's merchant fleet.[155] This could be encouraged by requiring a certain percentage of goods flowing into and out of American ports to be carried in American-built and American-crewed commercial ships. Given that the nation is already becoming a net exporter of energy and its manufacturing base is once again expanding, this would be wise. Any move to rebuild the nation's shipbuilding capacity must include a reinvestment in vocational training in our nation's public schools to enable them to help supply the welders, electricians, pipe fitters, and heating and air conditioning specialists that will be needed by shipyards and parts suppliers.

By extension, the nation also needs significant investments in the infrastructures of various port cities to make the import and export of goods most efficient and cost effective. An additional ancillary effect of this initiative would be the modernization of the rail lines and interstate highways that transport goods from the nation's interior to the ports and goods arriving in those ports from overseas into the nation's interior. A move to become a seapower once again would be transformative for the entire US economy.

Thus, the rebuilding and reordering of the American military toward a seapower approach to the world would become an extension of other reforms rather than a driver of them, and the result would be a broader, deeper, and more resilient industrial sector within the national economy, enabling it to withstand the shocks of war. The construction of naval ships, to include new, smaller unmanned craft, would begin in traditional shipyards along the Atlantic, Pacific, and Gulf coasts, but could quickly move inland to the Great Lakes as well as along the Mississippi and Ohio rivers, where smaller, shallower draft vessels were built in previous eras. Such a move would not only help to create much-needed high-skill and high-paying industrial jobs, it would also protect the nation against the danger of placing too many eggs — in terms of shipbuilding and the nation's broader industrial capacity — in a too few baskets, located along the nation's coastlines where facilities are imminently attackable by rising powers with long-range weapons. In the end, the country should field an overall military that is sized and shaped to compete with a rising China, and, to a lesser extent, Russia, in the oceanic domain where the United States both enjoys a present technological advantage and can exert influence events ashore without necessarily becoming entrapped by them.

Having adjusted the nation's strategic approach to the world and then rebuilt and strengthened its maritime economy, the nation will be positioned to expand and to change the US Navy to enable it to continue both to win the nation's wars and to preserve the global peace in a future competitive security environment. As such it will need a fleet of 450 ships, but it should be understood that if that at some future date, let us suggest 2040, it looks just like the Navy the nation has today, just bigger, then the nation has "missed the boat" and set itself up for failure.

Given that there is broad support for maintaining the supercarrier as the centerpiece of naval strike power, at least for now, then those carriers should be right-sized for the 65 aircraft air wings currently employed, which would allow for a smaller, cheaper carrier than the $15 billion *Ford* class currently under construction, and the embarked air wing should also be altered to again emphasize long-range penetrating strike as the air wings of the Cold War once did.[156] As a hedge against the possible increased vulnerability of the supercarrier, the Navy should increase its investment in submerged strike. The four *Ohio*-class guided missile submarines, each capable of launching 155 Tomahawk land attack missiles, have been known to change an enemy's calculations when they subtly reveal their silent and largely unseen presence in a region.[157] Unfortunately, these four ships and their large magazines are soon to retire from the fleet. An alternative evolutionary path could include the construction of eight new guided missile submarines in parallel with the eleven planned new *Columbia*-class nuclear ballistic missile submarines using the same basic hull design. Their capacity to generate massed strike must be maintained in the future force.

Finally, the fleet evolutionary pathway design needs more frigates to provide a cheaper and yet highly capable surface platform that can be purchased in sufficient numbers to meet the regional combat commander's day-to-day presence requirements and of such size and dimensions so as to operate alongside allied and partner navies and making port calls in the smaller confined ports that the US Navy has abandoned in recent decades. These frigates should be deployed with two or three medium unmanned surface vessels serving as sensor or shooter partners, extending the frigate's sensory range as well as deepening its magazine. The Marine Corps must also be encouraged to continue its own reforms and should be supported by a fleet of 40 light amphibious craft that will allow them access to the archipelagic waters where they plan to operate in the future. Taken together, all these requirements drive a requirement for a 456-ship fleet in 2040. Such a fleet would be a force that can preserve the concept of a free sea that the West has labored to build over the past four centuries.

	2020 Navy	2040 Navy
Carriers	11	11
Cruisers	22	22
Destroyers	70	73
Frigates	0	60
Corvettes	35	24
Large Unmanned Surface Vessels	0	18
Ballistic Missile Submarines	14	12
Guided Missile Submarines	4	8
Attack Submarines	50	62
Large Unmanned Underwater Vessels	0	21
Large Amphibious Ships	34	15
Small Amphibious Ships	0	40
Fleet Support Ships	55	90
Total Battleforce	295	456

The United States began its existence as a seapower state. The merchant, manufacturing New England states and the agricultural states of the south both understood that their prosperity lay in trade first with European, and later with Asian, markets. The nation's 1789 constitution indicated that the Congress would "raise and support armies," suggesting that the nation's continentally focused forces should be nonpermanent and optional, but the same section of the nation's founding charter obligates the Congress to "provide and maintain a navy," implicitly stating that the nation's maritime interests were enduring and permanent. However, events of the past seventy-five years have distracted the nation, lulling its people into believing that they could be all things to all peoples, that they could extend liberal democratic principles to all corners of the world, and that they could master all domains. Slowly, over the past decade, Americans have become aware of their limitations and begun to cast about for a new path forward. Fortunately, Americans are a resilient people, willing and able to reinvent themselves, and now is the time for such an action.

Given the preponderance of national interests that reside on and below the world's oceans, the continued favorable security conditions here in the American homeland, and the location and geographic distribution of the two rising great powers, it is time for the nation to fully invest in a return to the sea. It is the one strategic move the United States can make that can deter both Russia and China from future aggression today while also preparing to win a war tomorrow on its own terms. It is a move that will keep the world's oceans from once again becoming battlefields.

Acknowledgments

I would like to thank my good friends, advisors, and mentors Admiral James Stavridis, USN (retired), Rear Admiral Mark Montgomery, USN (retired), Hugh Hewitt, former Senator Jim Talent, and John Batchelor, all superb readers, who reviewed the initial drafts of this document and commented on its style and substance, improving the quality of the final product. In addition, I would like to thank my current and former co-workers, the late Shawn Brimley, Robert Work, Paul Scharre, Kelley Sayler, and Adam Routh at the Center for a New America Security, and Jim Thomas and Bob Martinage at the Telemus Group, who over the years listened to and debated the various arguments presented here, sharpening them. Many of them are still not convinced, but I like them anyway. I would also like to acknowledge my doctoral thesis advisor, Professor Andrew Lambert of Kings College, London, whose magisterial 2018 book *Seapower States* helped to frame this argument.

Of course, this book would not have been possible without the assistance of the team at Focsle Press, Claude Berube and Steve Phillips, along with their editors and production team at The Writer's Ally, who agreed to take this small book with big ideas, clean it up, and present it to the world. I must also acknowledge that had it not been for the COVID-19 virus, that horrible killer of so many wonderful people, and my forced isolation at home for the past nine months, I would not have had the time or concentration to finally bring all of these thoughts together in one place.

Finally, I want to thank my wife, Penny, my lifelong friend whose quick, brilliant mind has always been more than a match for me. Her constant support and analytical critique of my ideas during our daily morning coffee hour immeasurably helped these arguments come together in a presentable form. Of course, the thoughts contained are my own and do not necessarily reflect those of my employers or clients, and responsibility for any mistakes or inaccuracies rests solely with me.

Notes

1. Unless otherwise noted, all US Navy historical battleforce ship count numbers within this essay are derived from "US Ship Force Levels," Naval History and Heritage Command, U.S. Navy, https://www.history.navy.mil/research/histories/ship-histories/us-ship-force-levels.html#2000.

2. For a deeper exploration of the nature of naval presence, consider Jerry Hendrix, CDR B.J. Armstrong, "The Presence Problem," Center for New American Security, January 2016, https://s3.amazonaws.com/files.cnas.org/documents/The_Presence_Problem_FINAL.pdf?mtime=20160906082551.

3. "Share of the World Population living in Absolute Poverty, 1820–2015," OurWorldinData.org, accessed September 12, 2020.

4. Sun Tzu, *The Art of War* (New York: Delacorte Press, 1983), p. 16.

5. Halford Mackinder, "The Geographical Pivot of History," *Geographical Journal* 23, no. 4 (April 1904): 421–437.

6. Adam Smith, *The Wealth of Nations* (New York: Random House, 1937), pp. 398–464. David Ricardo, *On the Principles*

of Political Economy and Taxation, 3rd ed. (London: John Murray, Albemarle St., 1821), pp. 131–161.

7. Alfred T. Mahan, *Mahan on Naval Strategy,* edited by John B. Hattendorf (Annapolis, MD: Naval Institute Press, 1991), pp. 368–381.

8. Julian S. Corbett, *Some Principles of Maritime Strategy* (Annapolis, MD: Naval Institute Press, 1988), pp. 91–96.

9. Peter Garnsey and Richard Saller, *The Roman Empire: Economy, Society and Culture* (Los Angeles University of California Press, 1987), pp. 83–103.

10. David Brennan, "China Expands Military at India Border as Modi Accused of Surrendering Land, *Newsweek,* July 13, 2020, https://www.newsweek.com/china-expands-military-india-border-narendra-modi-accused-surrendering-land-1517300.

11. Eliot Cohen, "The Chinese Intervention in Korea, 1950," Central Intelligence Agency, approved for release 22 September 1993, https://www.cia.gov/library/readingroom/docs/1988-11-01.pdf.

12. Michael O'Hanlon, *The Future of Land Warfare* (Washington, DC: Brookings, 2015), p. 8.

13. Washington, George, "Farewell Address, September 17, 1796," *Messages and Papers of the Presidents,* vol. 1 (New York: Bureau of National Literature, 1897), pp. 214–215.

14. Theodore Sorenson, *Let the Word Go Forth: The Speeches, Statements and Writings of John F. Kennedy* (New York: Delacorte Press, 1988), p. 12. With regard to the covetous glances, there is Secretary of State Madeleine Albright's famous statement, "What's the point of having this superb military if we can't use it?" Colin Powell, *My American Journey* (New York: Random House, 1995), p. 576.

15. Tom Clancy and General Fred Franks Jr., *Into the Storm: A Study in Command* (New York: Putnam Books), 1997.

16. John Nagle, *Learning to Eat Soup with a Knife* (Chicago: University of Chicago Press, 2005). Janet Ritz, "Credit Where It is Due: HR McMaster, the Author of the "Surge" Strategy, *Huffington Post,* July 26, 2008, https://www.huffpost.com/entry/credit-where-it-is-due-hr_b_113202?.

17. Paul Scharre, *Army of None: Autonomous Weapons and the Future of War* (New York: WW Norton, 2018), pp. 59–100.

18. Christian Brose, *The Kill Chain: Defending America in the Future of High-Tech Warfare* (New York: Hachette Books, 2020), pp. 225–249.

19. Sydney J. Freedberg, Jr., "Army Targets AirSea Battle; Hungers for Pacific Role," *Breaking Defense,* December 13, 2011, https://breakingdefense.com/2011/12/army-targets-airsea-battle-hungers-for-pacific-role/.

20. "Factsheet: People and Oceans," Ocean Conference, United Nations, June 2017, https://www.un.org/sustainabledevelopment/wp-content/uploads/2017/05/Ocean-fact-sheet-package.pdf.

21. "World GDP," World Bank Group, https://data.worldbank.org/indicator/NY.GDP.MKTP.CD, accessed May 28, 2020. Jan Hoffmann, "Review of Maritime Transport 2018," *United Nations* (New York: United Nations Publications, 2018).

22. Robert E. Babe, "Sustainable Development vs Sustainable Ecosystem," *Culture of Ecology: Reconciling Economics and Environment* (Toronto: University of Toronto Press, 2006), pp. 3–29, http://www.jstor.org/stable/10.3138/9781442673663.5, accessed September 13, 2020.

23. Stephen D. Cohen, "General Agreement on Tariffs and Trade," in *Encyclopedia of U.S. Foreign Relations,* vol. 2, Bruce Jentleson and Thomas Paterson, eds. (Oxford, UK: Oxford University Press, 1997), pp. 203–207.

24. Ira Breskin, *The Business of Shipping,* 9th ed. (Atglen, PA: Cornell Maritime Press), pp. 196–197.

25. Ibid, pp. 310–344.

26. Vaclav Smith, *Prime Movers of Globalization: The History and Impact of Diesel Engines and Gas Turbines* (Cambridge, MA: MIT Press, 2010), pp. 109–127.

27. Martin Stopford, *Maritime Economics,* 3rd ed. (New York: Routledge, 2009), p. 43.

28. Ibid, p. 74.

29. William Glover, "History of the Atlantic Cable and Undersea Communications: Cable Timeline 1850–2018," February 25, 2019, https://atlantic-cable.com/Cables/CableTimeLine/index.htm.

30. Makada Henry-Nickie, Kwadwo Frimpong, and Hao Sun, "Trends in the Information Technology Sector," *Brookings,* March 29, 2019, https://www.brookings.edu/research/trends-in-the-information-technology-sector/

31. "Offshore Petroleum History," American Oil and Gas Historical Society, https://www.aoghs.org/offshore-history/offshore-oil-history/, accessed September 12, 2020.

32. "Deepwater Oil: Huge Growth Potential," *Seeking Alpha,* https://seekingalpha.com/article/3231866-deepwater-oil-huge-growth-potential, June 3, 2015.

33. Yong Bai and Quiang Bai, "Overview of Subsea Engineering," *Subsea Engineering Handbook* (Houston: Gulf Professional Publishing, Elsevier, 2019).

34. Rebecca McClay, "How the Oil and Gas Industry Works," *Investopedia,* March 6, 2020, https://www.investopedia.com/investing/oil-gas-industry-overview/.

35. Barclay Ballard, "Deep-sea mining could provide access to a wealth of valuable minerals," *New Economy,* May 13, 2019,

https://www.theneweconomy.com/energy/deep-sea-mining-could-provide-access-to-a-wealth-of-valuable-minerals.

36. Wil S. Hylton, "History's Largest Mining Operation Is About to Begin," *Atlantic*, January 2020, https://www.theatlantic.com/magazine/archive/2020/01/20000-feet-under-the-sea/603040/.

37. Carrie Donovan, "The Law of the Sea Treaty," *Heritage Foundation*, April 2, 2004, https://www.heritage.org/report/the-law-the-sea-treaty.

38. Max Roser and Esteban Ortiz-Ospina, "Global Extreme Poverty," *Our World in Data*, March 27, 2017, https://ourworldindata.org/extreme-poverty.

39. Max Roser and Esteban Ortiz-Ospina, "Literacy," *Our World in Data*, September 20, 2018, https://ourworldindata.org/literacy.

40. Max Roser, Esteban Ortiz-Ospina, and Hannah Ritchie, "Life Expectancy," *Our World in Data*, October 2019, https://ourworldindata.org/life-expectancy.

41. James Kraska and Raul Pedrozo, *The Free Sea: The American Fight for Freedom of Navigation* (Annapolis, MD: Naval Institute Press, 2018), pp. 3–4.

42. Hugo Grotius, *The Free Sea* (Indianapolis, IN: Liberty Fund, 2004; originally published 1609).

43. John Selden, *Mare Clausum of the Dominion, Or, Ownership of the Sea* (Clark, NJ: Law Book Exchange, 2014).

44. Cornelis van Bijnkershoek, *On Questions of Public Law* (London: Clarendon Press, 1930).

45. James Kraska and Raul Pedrozo, *The Free Sea* (Annapolis, MD: Naval Institute Press, 2018), p. 265.

46. Lawrence Mott, "Iberian Naval Power, 1000–1650," in *War at Sea in the Middle Ages and the Renaissance,* John Hattendorf

and Richard Unger, eds. (Suffolk, UK: Boydell Press, 2002), pp. 105–118.

47. Unless otherwise noted, all future references to the number of battles in a given period are derived from the following sources: Carl Ploetz, *An Outline of Universal History* (New York: Houghton Mifflin, 1914); H. G. Wells, *The Outline of History* (Garden City, NY: Garden City Books, 1949); P. H. Colomb, *Naval Warfare: Its Ruling Principles and Practice Historically Treated,* vol. 2 (Annapolis, MD: Naval Institute Press, 1990); E. B. Potter and Chester Nimitz, eds., *Sea Power, A Naval History* (Englewood Cliffs, NJ: Prentice-Hall, 1960); S. S. Robison and L. Mary, *A History of Naval Tactics from 1530 to 1930* (Annapolis, MD: Naval Institute Press, 1942); John Laffin, *Brassey's Battles: 3,500 Years of Conflict, Campaigns and Wars from A-Z* (New York, Brassey's Defence Publishers, 1986); R. G. Grant, *Battle at Sea: 3,000 Years of Naval Warfare* (New York: Dorling Kindersley Limited, 2008).

48. Jan Glete, *Navies and Nations: Warships, Navies and State Building in Europe and America, 1500–1860,* vol. 1, appendix 2 (Stockholm: Almqvist and Wiksell International, 1993).

49. E. B. Potter and Chester W. Nimitz, eds., *Sea Power* (Englewood Cliffs, NJ: Prentice-Hall, 1960), pp. 114–117.

50. Ibid.

51. Angus Maddison, *The World Economy: Historical Statistics* (Paris, Organization for Economic Cooperation and Development, 2003).

52. Nicholas Lambert, *Sir John Fisher's Naval Revolution* (Columbia, SC: University of South Carolina Press, 1999), p. 18.

53. Hans J. Morgenthau, *Politics Among Nations: The Struggle for Power and Peace* (New York: Alfred A. Knopf, 1961), p. 180.

54. Harold Nicolson, *Diplomacy* (London: Oxford University Press, 1960), p. 135.

55. William L. Langer, *European Alliances and Alignments* (New York: Alfred A. Knopf, 1956), pp. 1–19.

56. Charles Seymour, *The Diplomatic Background of the War: 1870–1914* (New Haven: Yale University Press, 1927), pp. 162–174.

57. William L. Langer, *The Diplomacy of Imperialism* (New York: Alfred A. Knopf, 1951, pp. 432–436.

58. Stephen M. Walt, *The Origins of Alliance* (Ithaca, NY: Cornell University Press, 1987), pp. 17–19.

59. Robert K. Massie, *Dreadnought: Britain, Germany, and the Coming of the Great War* (New York: Random House, 1991), pp. 181–183.

60. Michael Howard, *The Lessons of History* (New Haven: Yale University Press, 1991), pp. 81–96.

61. Chester Nimitz, and E. B. Potter, eds., *Sea Power* (Englewood Cliffs, NJ: Prentice Hall, 1960), pp. 432–454.

62. "The Royal Navy's Size Throughout History," https://www.historic-uk.com/Blog/British-Navy-Size-Over-Time/.

63. S. G. Gorshkov, *The Sea Power of the State* (Malabar, FL: Robert E. Krieger Publishing Co., 1979), p. 149.

64. Ibid. pp. 178–212.

65. North Atlantic, North Sea, Baltic Sea, Arctic, Eastern South Atlantic (Coast of Africa), Western South Atlantic (Coast of South of America), Mediterranean, Black Sea, Red Sea, Gulf of Oman, Arabian Gulf, Indian Ocean, South China Sea, East China Sea, archipelagic Pacific, North Eastern Pacific, South Eastern Pacific, Central Pacific, Antarctic.

66. Donald Daniel, "'Defense National' Perceptions of the US-Soviet Military Balance," Naval Postgraduate School, Monterey, CA, November 1976.

67. For analysis of the historical models of US Navy deployments, I am indebted to Captain Peter Swartz, USN (Ret) and in

particular his brief "U.S. Navy 'deployment strategies:' What are some alternatives?" February 3, 2011, Center for Naval Analysis.

68. Edward L. Beach, *The United States Navy* (New York, Henry Holt and Company, 1986), pp. 45–48, 342–347.

69. While some might count the United States' Operation Praying Mantis attack upon Iran's Navy on April 18, 1988, in retaliation for the near loss of the US frigate *Samuel B. Roberts* to an Iranian mine as a "war," this one-day action, despite its near destruction of the Iranian Navy, should not be counted as a single naval battle. "U.S. Strikes 2 Iranian Oil rigs and Hits 6 Warships in Battles over Mining Sea Lanes," *New York Times,* April 19, 1988, p. 1.

70. William Butler Yeats, "The Second Coming," *Poetry Foundation,* https://www.poetryfoundation.org/poems/43290/the-second-coming, accessed September 4, 2020.

71. Colleen Nash, "The Defense Budget — Revised," *Air Force Magazine,* July 1989, p. 67.

72. "Department of Defense Budget for FY 1999, Release No. 026-98," *Department of Defense,* February 2, 1998. https://archive.defense.gov/releases/release.aspx?releaseid=1566.

73. This was previously discussed with Julianne Smith in the report "Forgotten Waters: Minding the GIUK Gap," Center for a New American Security, May 2017, https://s3.amazonaws.com/files.cnas.org/documents/CNASReport-GIUKTTX-Final.pdf?mtime=20170502033816.

74. The broader topic of the growing divide between American and European perceptions of the security environment and the need to invest in an integrated defense infrastructure is covered brilliantly in Robert Kagan's *Of Paradise and Power: America and Europe in the New World Order* (New York: Alfred A. Knopf, 2003).

75. "Collision between US Navy Destroyer *John S McCain* and Tanker *Alnic MC* Singapore Strait, 5 Miles Northeast of Horsburgh Lighthouse August 21, 2017," National Transportation Safety Board, Marine Accident Report, NTSB/Mar-19/01, https://news.usni.org/2019/08/06/ntsb-accident-report-on-fatal-2017-uss-john-mccain-collision-off-singapore.

76. Sam Lagrone, "Repair for USS Fitzgerald After Collision Will Cost More Than to Fix to USS Cole After Terror Attack," USNI News, July 27, 2017, https://news.usni.org/2017/07/27/repair-bill-uss-fitzgerald-collision-will-cost-fix-uss-cole-terror-attack. "Repairing US Destroyer John S McCain to Cost $233 Million," DefenseWorld.Net, October 5, 2017, https://www.defenseworld.net/news/20858/Repairing_US_Destroyer_John_S_McCain_to_Cost__233_Million.

77. My thoughts on seapower and continental states have been greatly informed by my Kings College doctoral thesis director, Andrew Lambert, and his book, *Seapower States: Maritime Culture, Continental Empires and the Conflict that Made the Modern World* (New Haven: Yale University Press, 2018).

78. Zoltan Barany, "The Tragedy and Symbolism of the Kursk." In *Democratic Breakdown and the Decline of the Russian Military* (Princeton, NJ, Princeton University Press, 2007), pp. 19–43.

79. Frank Gibney, *The Pacific Century* (New York: Charles Scribner's Sons, 1992), pp. 290–293.

80. M. Taylor Fravel, *Active Defense* (Princeton: Princeton University Press, 2019), pp. 162–163.

81. Toshi Yoshihara and James R. Holmes, *Red Star Over the Pacific* (Annapolis, MD: Naval Institute Press, 2018), p. 109.

82. Bernard Cole, *Asian Maritime Strategies: Navigating Troubled Waters* (Annapolis, MD: Naval Institute Press, 2013) pp. 99–101.

83. Gregory J. Moore, "Not Very Material but Hardly Immaterial: China's Bombed Embassy and Sino-American Relations," *Foreign Policy Analysis* 6, no. 1 (January 2010): 23–41.

84. Andrew Krepinevich, Barry Watts, and Robert Work, *Meeting the Anti-Access and Area Denial Challenge* (Washington, DC: Center for Strategic and Budgetary Assessments, 2003).

85. Marshall Goldman, *Petrostate: Putin, Power and New Russia* (New York: Oxford University Press, 2010).

86. Pauline Davidson, "Kursk Disaster Still Haunts Russia's Naval Renewal," *World Review,* September 4, 2013, https://worldreview.info/content/kursk-disaster-still-haunts -russias-naval-renewal.

87. Charlie Bradley, "Putin's new Iron Curtain? How Russia is rebuilding Soviet legacy in Eastern Europe," *Express* 7, November 2019, https://www.express.co.uk/news/world/1201368/ putin-news-iron-curtain-russia-military-eastern-europe-bal-kan-states-eu-spt.

88. *"Little Green Men": Modern Russian Unconventional Warfare, Ukraine 2013–2014* (Fort Bragg, NC: US Army Special Operations Command), https://www.jhuapl.edu/Content/docu-ments/ARIS_LittleGreenMen.pdf.

89. Robert Mason, "Russia in Syria: An Unequivocal Return to the Middle East?" *Middle East Policy Council Journal* 25, no. 4 (Winter 2018), https://mepc.org/journal/ russia-syria-unequivocal-return-middle-east.

90. Paul Goble, "Russian Navy Ever Less Capable of Supporting Putin's War Plans," Jamestown Foundation, April 26, 2019, https://www.realcleardefense.com/articles/2019/04/26/rus-sian_navy_ever_less_capable_of_supporting_putins_war_ plans_114368.html.

91. Alice Scarsi, "Russia tensions SOAR as Putin's submarines 'in the Atlantic' — US warship DEPLOYED," *Express,* July 3,

2018, https://www.express.co.uk/news/world/982871/russia-tensions-us-vladimir-putin-submarines-uss-truman-deployed-atlantic.

92. Sydney J. Freedberg, Jr., "Outgunned Allies Must Contest Baltic, Black Seas: NATO Admiral, Breaking Defense," January 19, 2018, https://breakingdefense.com/2018/01/outgunned-allies-must-contest-baltic-black-seas-nato-admiral/.

93. This topic has been covered in deeper detail in "The United States Must Defend Open Seas in the Arctic," *National Review,* May 13, 2020, https://www.nationalreview.com/2020/05/arctic-us-navy-must-defend-open-seas/.

94. "How Is China's Energy Footprint Changing?" *China Power Project,* Center for Strategic and International Studies, https://chinapower.csis.org/energy-footprint/, accessed February 19, 2020.

95. John Fairbank, Edwin Reischauer, and Albert Craig, *East Asia: Tradition and Transformation,* rev. ed. (San Francisco: Wadsworth Publishing, 1989), pp. 440–441.

96. Mulvenon, James, "Chairman Hu and the PLA's "New Historic Missions," *China Leadership Monitor,* no. 27, Hoover Institute, https://media.hoover.org/sites/default/files/documents/CLM27JM.pdf.

97. Li, Nan, "The Evolution of China's Naval Strategy and Capabilities: From 'Near Coast' and 'Near Seas' to 'Far Seas,'" *Asian Security* 5, no. 2 (2009): 144–169.

98. Mark McDonald, "China's Navy to Join Pirate Patrols," *New York Times,* December 25, 2008, https://www.nytimes.com/2008/12/26/world/asia/26china.html.

99. Notes from personal conversation with Andrew W. Marshall, Director of the Office of Net Assessment within the Office of the Secretary of Defense, December 8, 2010.

100. Jessica Drun, "China's Maritime Ambitions: A Sinister String of Pearls or a Benevolent Silk Road (or Both)?" *Center for Advanced China Research,* December 5, 2017, https://www.ccpwatch.org/single-post/2017/12/05/China%E2%80%99s-Maritime-Ambitions-a-Sinister-String-of-Pearls-or-a-Benevolent-Silk-Road-or-Both.

101. Thomas Friedman, during an interview with Comedy Central's Stephen Colbert in November 2008, discussed his "fantasy" of the United States being "China for a Day" in order for the nation to make the series of decisions necessary to address the economic crisis of the Great Recession then ongoing, https://www.newsbusters.org/blogs/nb/tim-graham/2008/11/21/thomas-friedmans-power-lust-lets-be-china-day.

102. Jonathan Hillman, *The Emperor's New Road* (New Haven: Yale University Press, 2020).

103. Claude A. Buss, *The Far East* (New York, The Macmillan Co., 1955), pp. 66–71.

104. Richard Fisher, "China Buys New Russian Destroyers," *China Brief* 2, no. 3 (January 31, 2002), https://jamestown.org/program/china-buys-new-russian-destroyers/.

105. John Patch, "A Thoroughbred Ship-Killer," *US Naval Institute Proceedings,* April 2010, https://www.usni.org/magazines/proceedings/2010/april/thoroughbred-ship-killer.

106. Leszek Buszynski, "Chinese Naval Strategy, the United States, ASEAN and the South China Sea," *Security Challenges* 8, No. 2 (Winter 2012): 19–21.

107. Meyer, Manfred, *Modern Chinese Maritime Forces* (Washington, DC: Admiralty Trilogy Group, 2020), p. 12.

108. David Axe, "China's Giant New Warship Packs Killer Long-Range Missiles, *National Interest,* December 31, 2019, https://nationalinterest.org/blog/buzz/china%E2%80%99s-giant-new-warship-packs-killer-long-range-missiles-109786.

109. *Conway's Fighting Ships: 1906–1921* (London, UK: Conway Maritime Press, 1985), p. 67.

110. Thomas Hone, "Replacing Battleships with Aircraft Carriers in the Pacific in World War II," *Naval War College Review* 66, no. 1, pp. 56–76.

111. Raymond Cohen, "'Where are the Aircraft Carriers?': Nonverbal Communication in International Politics," *Review of International Studies* 7, no. 2, pp. 79–90.

112. Daniel J. Kostecka, "From the Sea: PLA Doctrine and the Employment of Sea-Based Airpower," *Naval War College Review* 64, no. 3, pp. 12–13.

113. Xiaoyu Pu, *Rebranding China: Contested Status Signaling in the Changing Global Order* (Stanford, CA: Stanford University Press, 2019), pp. 52–56.

114. Andrew Erickson and Lyle Goldstein, "China's Future Nuclear Submarine Force," *Naval War College Review* 60, no. 1, pp. 54–80.

115. Mike Yeo, "China completing more ballistic missile subs, with plans for a new version," *Defense News,* May 6, 2019, https://www.defensenews.com/global/asia-pacific/2019/05/06/china-completing-more-ballistic-missile-subs-with-plans-for-a-new-version/.

116. H. I. Sutton, "China's Submarine Force May Be Catching Up With US Navy," *Forbes,* November 24, 2019, https://www.forbes.com/sites/hisutton/2019/11/24/latest-chinese-submarines-catching-up-with-us-navy/#6a889248298c.

117. "Chinese Warships," GlobalSecurity.Org, https://www.globalsecurity.org/military/world/china/navy.htm., accessed February 11, 2020.

118. Admiral Harry Harris, US Navy, first described China's artificial and illegal islands in the South China Sea as a "great wall of sand" in April 2015. Harris was then serving as the

Commander of the US Pacific Command. "China building 'great wall of sand' in South China Sea, BBC News, April 1, 2015, https://www.bbc.com/news/world-asia-32126840.

119. "Occupation and Island Tracker," Asia Maritime Transparency Initiative, Center for Strategic and International Studies, https://amti.csis.org/island-tracker/, accessed February 12, 2020.

120. Keith Bradsher, "Philippines Challenges China Over Disputed Atoll," *New York Times,* May 14, 2014, https://www.nytimes.com/2014/05/15/world/asia/philippines-challenges-china-over-disputed-atoll.html.

121. Mobo Gao, *Constructing China* (London, UK: Pluto Press, Inc., 2018), pp. 227–231.

122. Euan Graham, "China's New Map: Just Another Dash?" *RUSI News Brief,* September 3, 2013, http://www.rusi.org/publications/newsbrief/ref:A5225D72CD72F8/.

123. Jane Perlez, "Philippines v. China: Q and A on South China Sea Case," New York Times, July 11, 2016, https://www.nytimes.com/2016/07/11/world/asia/south-china-sea-philippines-hague.html.

124. The Honorable Michael Pompeo, US Secretary of State, "U.S. Position on Maritime Claims in the South China Sea," US State Department, July 13, 2020, https://www.state.gov/u-s-position-on-maritime-claims-in-the-south-china-sea/

125. Thomas Shugart, "China's Artificial Islands are Bigger and a Bigger Deal Than You Think," WarOnTheRocks.Com, September 21, 2016, https://warontherocks.com/2016/09/chinas-artificial-islands-are-bigger-and-a-bigger-deal-than-you-think/.

126. Andrew Erickson and David Yang, "On the Verge of a Game Changer," *Naval Institute Proceedings* 135, no. 5, pp. 26–32.

127. Sam Tangredi, "Fight Fire with Fire," *Naval Institute Proceedings* 143, no. 8, pp. 42–47.

128. "A Cooperative Strategy for the 21st Century," US Navy, October 2007, p. 4.

129. "10 U.S. Code Section 8062," https://www.law.cornell.edu/uscode/text/10/8062.

130. George W. Baer, *One Hundred Years of Sea Power* (Stanford, CA: Stanford University Press, 1994), p. 430.

131. Norman Friedman, *U.S. Battleships: An Illustrated Design History* (Annapolis, MD: Naval Institute Press, 1985), p. 420.

132. Norman Friedman, *U.S. Aircraft Carriers, An Illustrated Design History* (Annapolis, MD: Naval Institute Press, 1983), pp. 255–271.

133. This argument was presented in a longer form in my report "Retreat from Range: The Rise and Fall of Carrier Aviation," Center for a New American Security, October 2015.

134. This argument was advanced in my report "Filling the Seams in U.S. Long-Range Penetrating Strike," Center for a New American Security, September 2018.

135. Bernard Brodie, *A Layman's Guide to Naval Strategy* (Princeton, NJ: Princeton University Press, 1942), pp. 45–47.

136. I have presented this argument in longer form as "Building a 355-Ship Navy: It's Not Just the Number, It's the Mix," in *National Interest,* September 25, 2017.

137. David Larter, "All US Navy Destroyers will get Hypersonic Missiles, Trump National Security Advisor Says," Defense News, Oct 21, 2020, https://www.defensenews.com/naval/2020/10/21/all-us-navy-destroyers-will-get-hypersonic-missiles-trumps-national-security-advisor-says/.

138. I have presented this argument in longer form as "The Navy We Need," in *National Review,* November 12, 2018, pp. 27–30.

139. James Q. Wilson and George L. Kelling, "Broken Windows," *Atlantic,* March 1982.

140. Tom Phillips, Oliver Holmes, and Owen Bowcott, "Beijing Rejects Tribunal's Ruling in South China Sea Case," *Guardian,* July 12, 2016, https://www.theguardian.com/world/2016/jul/12/philippines-wins-south-china-sea-case-against-chia.

141. "Piracy, other high seas crimes rise in Asia — report," *Yahoo News,* September 9, 2020, https://news.yahoo.com/piracy-other-high-seas-crimes-031024978.html. Captain James Fanell, USN (Ret), "Comments on the South China Sea," July 7, 2020, provided to the author. Captain Fanell is a former senior Navy intelligence officer with tours on the Pacific Fleet and Pacific Command staffs during the period under consideration.

142. This is covered in greater detail in "Iran's Arrest of U.S. Sailors Reflects Obama's Foreign Policy Weakness," *National Review,* January 18, 2018, https://www.nationalreview.com/2016/01/iran-united-states-navy-sailors-obama-foreign-policy/.

143. David Larter, "US Navy Proposes Decommissioning First 4 LCS More than a Decade Early," *Defense News,* December 24, 2019, https://www.defensenews.com/naval/2019/12/24/us-navy-proposes-decommissioning-first-4-lcs-more-than-a-decade-early/.

144. Hugh Hewitt, "How Trump Can Rebuild the Navy, Create Job-and Send China a Message," *Washington Post,* May 3, 2020, https://www.washingtonpost.com/opinions/2020/05/03/how-trump-can-rebuild-navy-create-jobs-send-china-message/.

145. This argument was previously presented as "Buy Fords, Not Ferraris," in *Naval Institute Proceedings,* April 2009.

146. Megan Eckstein, "Marines Look to Two New Ship Classes to Define Future of Amphibious Operations, *USNI News,* June 12, 2020, https://news.usni.org/2020/06/08/marines-look-to-two-new-ship-classes-to-define-future-of-amphibious-operations.

147. This argument was previously presented in "More Henderson, Less Bonds," Naval Institute Proceedings, April 2010, pp. 60–65.

148. Michael Lewis, *Moneyball* (New York: W.W. Norton & Co., 2003), pp. 119–137.

149. Rep. Jim Banks, H. Res. 99, 116th Congress, 1st Session, https://banks.house.gov/uploadedfiles/five_ocean_navy_strategy_-_official.pdf.

    "House Armed Services Committee Holds Hearing on the Proposed Fiscal 2015 Defense Authorization for the Navy Department, Greenert Testimony," 12 March 2014.

150. "Budget of the US Navy: 1794–2014," Naval History and Heritage Command, https://www.history.navy.mil/research/library/online-reading-room/title-list-alphabetically/b/budget-of-the-us-navy-1794-to-2004.html, accessed August 7, 2020.

151. Niall McCarthy, "The Biggest Military Budgets as a Share of GDP in 2019," *Forbes,* April 27, 2020, https://www.forbes.com/sites/niallmccarthy/2020/04/27/the-biggest-military-budgets-as-a-share-of-gdp-in-2019-infographic/#f2cfc1137f10.

152. The words have often been attributed to Sir Winston Churchill, but research has been unable to find him actually uttering them. However, a derivation of them was said by the Israeli politician Abba Eban in 1967, and they have been carefully altered and matured ever since, https://quoteinvestigator.com/2012/11/11/exhaust-alternatives/.

153. "National Security Strategy of the United States of America," The White House, December 2017.

154. "Summary of the 2018 National Defense Strategy of the United States of America: Sharpening the American Military's Competitive Edge," US Department of Defense, January 2018.

155. Arthur Herman, *Freedom's Forge* (New York: Random House, 2012), p. ix.

156. This argument was previously presented in longer form as "The Aircraft Carrier We Need," in *National Review,* June 18, 2020. https://www.nationalreview.com/magazine/2020/07/06/the-aircraft-carrier-we-need/#slide-1.

157. "North Korea tensions: USS Michigan submarine to enter South Korean port," Fox News, April 25, 2017, https://www.foxnews.com/world/north-korea-tensions-uss-michigan-submarine-to-enter-south-korean-port.

Printed in the USA
CPSIA information can be obtained
at www.ICGtesting.com
LVHW021409171123
764237LV00027B/349/J

9 780960 039197